Connections Management Strategies
in Satellite Cellular Networks

Series Editor
Guy Pujolle

Connections Management Strategies in Satellite Cellular Networks

Malek Benslama
Wassila Kiamouche
Hadj Batatia

WILEY

First published 2015 in Great Britain and the United States by ISTE Ltd and John Wiley & Sons, Inc.

Apart from any fair dealing for the purposes of research or private study, or criticism or review, as permitted under the Copyright, Designs and Patents Act 1988, this publication may only be reproduced, stored or transmitted, in any form or by any means, with the prior permission in writing of the publishers, or in the case of reprographic reproduction in accordance with the terms and licenses issued by the CLA. Enquiries concerning reproduction outside these terms should be sent to the publishers at the undermentioned address:

ISTE Ltd
27-37 St George's Road
London SW19 4EU
UK

www.iste.co.uk

John Wiley & Sons, Inc.
111 River Street
Hoboken, NJ 07030
USA

www.wiley.com

© ISTE Ltd 2015

The rights of Malek Benslama, Wassila Kiamouche and Hadj Batatia to be identified as the authors of this work have been asserted by them in accordance with the Copyright, Designs and Patents Act 1988.

Library of Congress Control Number: 2014956812

British Library Cataloguing-in-Publication Data
A CIP record for this book is available from the British Library
ISBN 978-1-84821-775-1

Contents

PREFACE. ix

ABBREVIATIONS xi

INTRODUCTION. xv

CHAPTER 1. THE FOUNDATIONS OF
SATELLITE NETWORKS. 1

  1.1. Introduction 1
  1.2. Satellite orbits. 3
    1.2.1. Characteristics of the ellipse 3
    1.2.2. Kepler's laws. 4
    1.2.3. Orbital parameters for earth satellites. 5
    1.2.4. Orbital perturbations 7
    1.2.5. Maintaining and surviving an orbit 7
  1.3. Time, time variation and coverage. 8
    1.3.1. Geometric data 8
    1.3.2. Approximation of coverage 11
    1.3.3. Time interval between two
    successive intersatellite transfers 12
    1.3.4. Time and time variation. 12
  1.4. Orbital paths. 13
    1.4.1. GEO-type systems. 14
    1.4.2. Elliptical systems 15
    1.4.3. MEO-type systems. 17
    1.4.4. LEO-type systems 17
  1.5. Characteristics of cellular satellite systems. 19

1.6. The advantages of LEO systems 22
1.7. Handover in LEO satellite networks. 23
  1.7.1. Link-layer handover 24
  1.7.2. Network-layer handover 25

CHAPTER 2. AN INTRODUCTION TO TELETRAFFIC 27

2.1. Introduction 27
2.2. The history of teletraffic theory and technique 28
  2.2.1. Queuing theory 28
  2.2.2. Teletraffic theory 29
2.3. Basic concepts 30
  2.3.1. The birth–death process 31
  2.3.2. Poisson process 32
2.4. Erlang-B and Erlang-C models 34
  2.4.1. Blocking probability and the Erlang-B formula. 34
  2.4.2. Queuing probability and the Erlang-C formula 36

CHAPTER 3. CHANNEL ALLOCATION STRATEGIES AND THE MOBILITY MODEL 39

3.1. Introduction 39
3.2. Channel allocation techniques 40
  3.2.1. Fixed channel allocation techniques 41
  3.2.2. Dynamic channel allocation techniques 41
3.3. Spotbeam handover and priority strategies 43
  3.3.1. Spotbeam handover 43
  3.3.2. Priority strategies for handover requests 45
3.4. Mobility model 48
3.5. Analysis of the mobility model 53

CHAPTER 4. EVALUATION PARAMETERS METHOD 63

4.1. Introduction 63
4.2. The advantages of the LEO MSS mobility model 64
4.3. Evaluation parameters method 71
  4.3.1. Position of the MU in the cell 71
  4.3.2. The moment the next handover request initializes 72
  4.3.3. Maximum queuing time 74

4.4. Pseudo-last useful instant queuing strategy.		77
4.4.1. Putting handover requests in a queue.		77
4.4.2. Handover request management		77
4.4.3. LUI queuing strategy		78
4.4.4. Pseudo-LUI queuing strategy		79
4.5. Guard channel strategy: dynamic channel reservation-*like*.		81
4.5.1. Dynamic channel reservation technique.		81
4.5.2. Dynamic channel reservation DCR-like technique.		83
CHAPTER 5. ANALYTICAL STUDY.		85
5.1. Introduction		85
5.2. An analysis of FCA-QH with different queuing strategies		85
5.3. Analytical study of FCR and FCR-*like*		91
5.3.1. An analysis of FCR		91
5.3.2. An analysis of FCR-like		94
CHAPTER 6. THE RESCUING SYSTEM		101
6.1. Introduction		101
6.2. Fuzzy logic.		102
6.2.1. Definition of fuzzy subsets.		102
6.2.2. Decisions in the fuzzy environment.		102
6.3. The problem		103
6.4. Rescuing system		105
CHAPTER 7. RESULTS AND SIMULATION		109
7.1. Introduction		109
7.2. The (folded) simulated network		110
7.3. Simulation results.		112
7.3.1. Verifying the simulation: a comparison with the analytical results of the FCA-QH case with different queuing strategies		113
7.3.2. A comparison of FCA and DCA, DCA-QH & FCA-QH simulation using LUI		115
7.3.3. A comparison of NPS and QH, DCA-NPS & DCA-QH simulation.		116

7.3.4. Comparison of QH strategies,
DCA-QH FIFO, LUI, PLUI simulation 117
7.3.5. Verifying the simulation: a
comparison with the analytical results
of the FCR and FCR-like case. 119
7.3.6. A comparison of DCR
and DCR-like. 120

**CHAPTER 8. PAB FOR IP TRAFFIC
IN SATELLITE NETWORKS** 127

8.1. Introduction 127
8.2. Proportional allocation of bandwidth 129
   8.2.1. Implementation of PAB. 130
8.3. Determination of the label fraction 135
   8.3.1. Equal fractions 135
   8.3.2. AP fractions. 135
   8.3.3. GP fractions. 135
8.4. Simulation and results 136
   8.4.1. Single congested link 137
   8.4.2. Multiple congested link 146
8.5. Conclusion 149

GENERAL CONCLUSION 151

APPENDIX 1 157

APPENDIX 2 161

APPENDIX 3 163

APPENDIX 4 167

APPENDIX 5 169

BIBLIOGRAPHY 181

INDEX 201

Preface

Four books devoted solely to satellite communication: this was the challenge laid down by Professor Malek Benslama of the University of Constantine, who understood that a new discipline was in the process of taking shape.

He demonstrated this by organizing the first International Symposium on Electromagnetism, Satellites and Cryptography in Jijel, Algeria, in June 2005. The success of this conference, which was surprising for an inaugural event, demonstrated the need for specialists with skills that sometimes varied widely from one another to come together in the same place. The 140 papers accepted concerned not only systems but also electromagnetism, antenna and circuit engineering, and cryptography, which often falls under the category of pure mathematics. Synergy must exist among these disciplines in order to develop the new field of activity that is satellite communication.

We have seen that new disciplines of this type emerge in the past; for electromagnetic compatibility, it was necessary to understand both electrical engineering (for guided modes and choppers) and electromagnetism (for propagated modes) and to know how to define specific experimental protocols as well. Further back in time, computer science was the domain of electronics engineers in its early days, and became a separate discipline only gradually.

Professor Benslama has the knowledge and open-mindedness needed to combine all the areas of expertise that coexist in satellite telecommunications. I have known him for 28 years now, and it has been a real pleasure for me to look back on all those years of acquaintance. Not a single year has gone by when we have not seen each other. He spent the first 15 years of his career working on the interaction between acoustic waves and semiconductors, specializing in the solution of piezoelectric equations (Rayleigh waves, surface skimming waves, etc.) while taking an interest in theoretical physics at the same time. A PhD degree in engineering and, later, a high-level State doctorate degree were added to his many achievements. Among the members of his dissertation committee was Madame Hennaf, then Chief Engineer at CNET (the National Centre for Telecommunications Studies in Issy Les Moulineaux). He had already developed an interest not only in telecommunications but also, with the presence of Monsieur Michel Planat, head of research at the National Centre for Scientific Research at LPMO Besançon (CNRS), in the difficult problem of the synchronization of oscillators.

With Michel Planat, he embarked on the path that would lead him to quantum cryptography, a conversion that he has made over the past 10 years, passing without apparent difficulty from Maxwell equations to Galois groups.

He is now one of the people most capable of mastering all the diverse disciplines that form satellite telecommunications.

I hope, with friendly admiration, that these four monographs will receive a warm welcome from both students and instructors.

Professor Henri BAUDRAND
Professor Emeritus
ENSEEIHT
Toulouse
France
December 2014

Abbreviations

AP	Arithmetic progression
CIR	Committed information rate
CRN	Channel reservation number
DCA	Dynamic channel allocation
DCR	Dynamic channel reservation
DDBHP	Dynamic Doppler-based handover prioritization
DiffServ	Differentiated services
ECL	Elastic channel locking
EF	Equal fractions
EPM	Evaluation parameters method
FCA	Fixed channel allocation
FCR	Fixed channel reservation
FIFO	First in first out
GEO	Geostationary Earth orbit

GH	Guaranteed handover
GP	Geometric progression
GPS	Global positioning system
HEO	High Earth orbit
HG	Handover guard
IP	Internet Protocol
ISL	Intersatellite link
ITU	International Telecommunication Union
LEO	Low Earth orbit
LUI	Last useful instant
MBPS	Measurement based priority scheme
MEO	Medium Earth orbit
MLTQ	Multilevel based queuing
MSS	Mobile satellite system
MU	Mobile user
NPS	Non Prioritization Scheme
PAB	Proportional Allocation Bandwidth
PASTA	Poisson Arrivals See Time Averages
pdf	Probability density function
PLUI	Pseudo LUI
QH	Queuing handover
QoS	Quality of Service

RED	Random early discard
SR	Satellite router
TCP	Transmission Control Protocol
TCRA	Time-based channel reservation algorithm
TR1	Terrestrial Router 1
UDP	User Datagram Protocol
UIT	Union International Telecommunication

General Introduction

A major concern in telecommunications has of late been creating the most efficient methods for managing telecommunications networks. Efficient management is crucial in *ad hoc* networks [AKR 06], wireless networks [CHE 14] and cellular networks [VEE 14]. Managing the energetic power in the nodes of wireless networks has required a whole prediction process [PEN 14], while managing mobility in wireless networks is based on fuzzy logic [ZIN 14]. In large bandwidth low earth orbit (LEO) satellite networks, a specific router based on multi-service agents has been researched [RAO 14] that has particular requirements with regard to quality of service [YIN 09], namely the allocation strategy in networks [KAR 14], the optimization of resources [WAN 14] and interference [UYA 14]. These factors must be taken into consideration in every management process, given that access to the spectrum is shared in LEO satellite networks [XIE 12].

To attain efficient cellular management, projects have been developed to meet these pressing needs and to create a global network that is capable of providing radio access to users wherever they may be, using a universal personal address.

As the demand for a large range of telecommunication services (for instance, voice transmission, short messaging services and global positioning system (GPS) localization) continues to grow, radio access solutions appear increasingly attractive as they free users from attachment restrictions [PRA 05, MCC 07, DEL 08]. Projects have been developed to meet these pressing needs and to create a global network capable of providing radio access to users wherever they may be, using a universal personal address. These projects have become patents, which we have cited for readers' information [AUV 96, DIE 96, DEN 01, LYN 02].

The challenge is thus to implement universal communication services [WYS 05].

Mobile satellite systems (MSS) play a very important role in this challenge. These systems are expected to develop around their capacity to provide global coverage for a range of users on the earth, in the air or at sea. Moreover, MSS present a rare opportunity to guarantee communication services in sparsely populated areas or vast regions where implementation of a land mobile network is impossible or too costly (e.g. the ocean).

Recently, satellite network operators have become interested in LEO satellite constellations in view of the advantages they present in comparison to high earth orbit satellite constellations [JAM 97]. On the one hand, in good conditions, LEO satellites can communicate with very high latitudes and the polar areas, unlike geosynchronous satellites above the equator. On the other hand, their low altitude enables them to communicate with ground stations that are not very powerful or that have small antenna [ELZ 05]. A renewed interest in this type of satellite is enabling communications between mobiles and portable stations [SHE 01] to be established.

The end of the 20th Century was characterized by satellite constellations' hour of glory [GRU 91]. Indeed, in the final decade of the century, a number of commercial satellite constellations networks were constructed and implemented: Iridium 1998, Global Star 1998 and Teledesic 2002 [GRU 91, SUN 05, MON 05].

The greatest problem with regard to the design of these systems is establishing the best technique for communications routing that will maintain a reasonable quality of service [ZHA 03].

However, these systems face a major problem. The geometry of this type of constellation is dynamic and, given that the satellites move around quickly in comparison to ground stations, has short links. To avoid losing an established call, it is necessary to switch between one satellite (or spotbeam) and another if the user leaves the zone covered by the satellite (or the spotbeam). This transfer, commonly known as "handover", is an essential parameter for ensuring a good quality of service [AKY 99, CHO 06, DEL 95]. Indeed, the probability of a failed handover is a common criterion for evaluating the performance of satellite networks.

Although different types of handover in LEO MSS networks exist (handover between spotbeams, between satellites, etc.), handover between spotbeams has attracted a great deal of attention and has been the subject of extensive research because it occurs frequently. Indeed during an ongoing call, a handover request between spotbeams is expected every minute and sometimes even more frequently. This book will research handover between spotbeams.

From the perspective of phone users, the termination of an ongoing call is much less desirable than the rejection of an attempt to make a new call [TEK 92]. Consequently, several methods that give priority of service to handover requests and thus improve cellular quality of service have been presented in the literature [DEL 99, OBR 99, RAP 93, LIN 94, ZHA 04]. These methods or procedures can be broadly divided into two groups according to the following concepts:

– guaranteed handover systems;

– guaranteed priority systems for handover.

The first systems guarantee that handover requests will be successful: the probability of a handover request failing is thus nil. The other systems give priority of service to handover requests over new calls, thereby reducing the probability of a handover request failing at the expense of a sometimes tolerable rise of the probability of preventing new calls [SHE 96, HOM 98].

In systems that do not have an order of priority, handover requests are treated in the same way as new calls. The probability of losing a call during a handover procedure is therefore increased.

In [MAR 91], the coauthors propose a *guaranteed handover* (GH) system in which each call transferred toward a cell starts a request to reserve a channel in the next transit cell. This strategy reduces the probability of losing ongoing calls to values of almost zero at the expense of the probability of blocking new calls, which reached values that were much too high [CHA 01, LIN 94, LIA 02, BOU 02, DEL 96, SAN 95, CHO 06, MAR 98]. To improve the allocation of resources, some studies presented modified GH systems, for instance *elastic channel locking* (ECL) [XU 00], *time-based channel reservation algorithm* (TCRA) [BOU 03a, BOU 01, BOU 03b] and *dynamic Doppler-based handover prioritization* (DDBHP) technique [PAP 03, PAP 04].

Systems giving priority to handover requests can be divided into four groups according to the concept they adopt:

– handover with guard channels: the system reserves a number of channels exclusively for handover requests;

– handover with a queuing system: the system exploits the overlap zone between cells (in this zone the mobile user can be served by two cells) to put handover requests in a queue for a free channel in the transit cell for a given period;

– handover based on rearranging channels: only used with the dynamic allocation of channels [TEK 92]. When each call is terminated, the system rearranges the channels to liberate the channel that will become available in the maximum number of adjacent cells;

– handover with guard channels and queues: combines the advantages of using guard channels and handover queues.

In addition to the advantages that LEO satellite systems present for universal mobile telecommunications systems, such as their relatively low transmission power and short transmission time [JAM 01, GAN 94], another equally important advantage of these systems is that the relative movement between the mobile user (MU) and the satellite can be predicted. Indeed, the speed of the footprint the satellite casts on the ground is far higher than that of the MU and the rotation of the earth [RES 95]; the relative movement between the MU and the satellite can thus only be approximated by the movement of the satellite [ZHA 95].

This advantage enables the system to determine various important parameters, such as the maximum time a handover request can be in a queue and the moment the next handover request for an ongoing call starts. In research previously published in the literature, these parameters were determined by the fact that the system estimates the position of the MU at the beginning of each call. This requires a *positioning system* to be integrated into the LEO MSS [WAN 01, WAN 02, KOL 02].

This book will demonstrate that it is possible to determine some information and important parameters even if the exact location of an MU is unknown by exploiting the predictable behavior of the relative movement between the MU and the satellite and the regular position of the cells in the network.

Indeed, the fixed direction and the speed of the movement, in addition to the regular cell pattern, enable different parameters and information to be determined according to the maximum time it takes a MU with on ongoing call to pass through a cell, which is equal to the period separating two consecutive handover request initiations.

We are proposing a method, i.e. the *evaluation parameters method* (EPM), based on this idea, which we then use to introduce two priority strategies for handover requests [KIA 09].

We will first look into the concept of handover requests queues, and more specifically to the *last useful instant* (LUI) queuing strategy presented by Del Re *et al.* [DEL 99]. This strategy is seen as ideal as it is based on a precise estimation of the maximum period of time a call is in the overlap zone and thus in a queue for a free channel. In comparison with the most popular queuing strategy, *first in first out* (FIFO) [DEL 95a, DEL 95b, HON 86], LUI guarantees a large system capacity as it always serves the most urgent handover request. It is, however, more complicated to implement. Indeed, a *positioning system* is required to estimate the position of the MU starting a call and to follow him or her while the call is ongoing [DEL 99, MAR 98, ZHA 95].

To reach a compromise between these two strategies, we will put forward a new queuing strategy that combines the FIFO's ease of implementation and the LUI's efficacy. This strategy, called pseudo-LUI, is essentially based on the EPM method that we are presenting in this research [KIA 08].

We will then look into the concept of priority, which affects guard channels for handover requests, and more specifically to *dynamic channel reservation* (DCR) presented by Wang *et al.* [WAN 01]. In this strategy, the number of guard channels is dynamically calculated by the *channel number reservation* (CNR) parameter system using several parameters concerning MUs with ongoing calls, including the position and the maximum time spent in the cell. It also requires a positioning system to track MUs with ongoing calls.

Based on the EPM method and the DCR strategy, the DCR-like [KIA 11] strategy that we will present is a combination of the *locking-channel mechanism* [MAR 98] and the DCR strategy.

We will also tackle another aspect of the handover problem: short calls that are lost due to a failure in the handover procedure. Indeed, the great mobility of LEO MSS systems results in short calls being interrupted a short while after they start. To resolve this problem, we will present a *rescuing system* (RS) that avoids these calls ending prematurely at the expense of relatively long calls.

Chapter 1 will discuss satellite networks, their characteristics, their advantages and the different types of handover procedures they encounter. Chapter 2 will present the history of queuing and teletraffic theory which are the basic tools for studying the performance of telephone communication systems.

Chapter 3 will then present the strategies affecting channels, several management strategies for handover requests and the mobility model we adopted in our research.

In Chapter 4, a new method for evaluating important EPM parameters will be presented, which we will then use to introduce two new handover priority strategies, pseudo-LUI and DCR-like, which are based on LUI and DCR strategies, respectively.

A theoretical study of various handover priority strategies will be given in Chapter 5, focusing particularly on the strategies proposed.

The RS system will be presented in Chapter 6. This last chapter is dedicated to listing and discussing the results obtained by different strategies.

During our research into cellular networks and in light of recent developments in the literature, it seemed the perfect time to introduce all the possible applications of Voronoi cells into wireless networks [FAN 04, CHO 09, FLE 06, WAN 07, XIE 07]. To conclude this book, we will provide an interesting application concerning the proportional allocation of bandwidth for IP traffic in satellite networks [JAN 03, LI 07].

1

The Foundations of Satellite Networks

1.1. Introduction

Satellite networks are principally characterized by the use of a satellite. They were designed with the aim of achieving a global coverage of the earth and to provide different services, such as voice transmission, short messaging services, global positioning system (GPS) localization, etc. These networks also enable two distant points located anywhere on the earth to communicate. However, building a network always requires that a number of precautions be taken and, more specifically, that a precise set of specifications relating to performance be presented. Theoretically, these specifications are determined from market research that aims to forecast the different uses of the system. It would be very naïve to think, however, that this research could forecast the future needs of users with any degree of certainty.

The market can change very quickly and a property that was overlooked when the network was being designed can become the determining factor in getting one step ahead of the competition. In this event, it is imperative that the network follows the change.

For example, let us consider land mobile networks, which have undergone countless changes in infrastructure. At first operators were attached to the idea of covering the largest possible area in a populated zone. Then, however, the market grew exponentially and operators refocused their efforts on regions that already had a wide coverage for capacity reasons: in certain situations, a station was no longer able to meet the demand from the zone it was covering, a zone that is called a *cell*. An effort was then made to divide the overloaded cells into smaller cells that were associated with less powerful (and therefore less *noisy*) stations, thereby enabling radio channels to be used once again.

The idea of using more, smaller and less powerful cells can also be applied to satellite networks. To increase the capacity of a system, satellites can be *brought closer* to the earth, thus improving acuity for differentiating between ground terminals, as illustrated in Figure 1.1: to cover an equivalent air surface, a satellite that is far from earth must use a much more precise spotbeam than a satellite that is closer to earth.

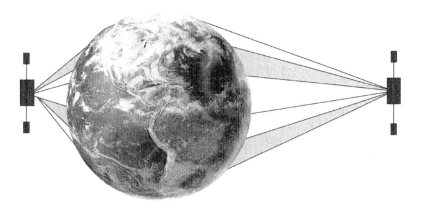

Figure 1.1. *The reuse of frequencies by satellites with varying altitudes*

This chapter will first present the places where a telecommunications satellite can be positioned in the vicinity of the earth. We will then see how several satellites can work together to guarantee a global coverage. We will also tackle the question of the time it takes to pass through a network according to where it is located and, in concrete terms, we will discuss the resulting broad categories of satellite networks. The characteristics of cellular satellite networks will then be presented and the problem of handover in LEO satellite networks will be introduced.

1.2. Satellite orbits

An orbit is a path the satellite follows when there are no perturbations. There are many different types of satellite orbits. Contrary to common opinion, a device's path without propulsion has nothing at all to do with its weight, though it is conditioned by precise rules (described next). Before this, it is useful to go over some basic properties of the ellipse.

1.2.1. *Characteristics of the ellipse*

Let us consider a plane with an orthonormal coordinate system $(O, \vec{i}, \vec{j})$. Given the two real positive constants a and b where $a > b$, the ellipse ξ centered on O of a semi major axis a and a semi minor axis b is the set of points $P(x, y)$ that verify the following equation:

$$\frac{x^2}{a^2} + \frac{y^2}{b^2} = 1 \qquad [1.1]$$

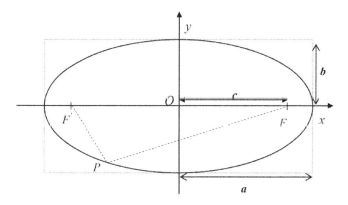

Figure 1.2. *A few properties of the ellipse*

A general parametric form of the ellipse exists. It can be formulated as follows:

$$\varepsilon = \left\{ P(x,y) \quad with \begin{vmatrix} x = a\cos(k) \\ y = b\sin(k) \end{vmatrix} \right\} \quad with \quad k \in \mathrm{IR} \quad [1.2]$$

Figure 1.2 shows different parameters. The two foci of the ellipse appear at points $F(c, 0)$ and $F'(-c, 0)$ with $c = \sqrt{a^2 - b^2}$. The *eccentricity* of the ellipse is thus defined by $e = c / a$. For every point P of ε, the following relation is verified:

$$PF + PF' = 2a$$

1.2.2. *Kepler's laws*

The parameters used today to describe satellite orbits are inspired by the work of Kepler (1571–1630) [SUN 05, MON 05].

These laws describe the way in which planets move around the sun and are typical of the following parameters:

– the *orbit of planets*, that is the path they follow over time;

– the *instantaneous speed* of the path of each orbit by the associated planet;

– the *orbital period* of a planet, which is the total time it takes to complete its orbit;

Kepler's laws can be summarized as follows:

– the orbit of each planet is an ellipse where the sun is one of the foci;

– the area swept out by a line between the sun and the planet is constant over a unit of time;

– the square of an orbital period of a planet is proportional to the cube of the major axis of its orbit.

These laws can, in fact, be applied to any system where a celestial body with a large mass (the sun), which is known as the *primary element*, determines the movement of bodies with small masses in comparison to the former (the planets), which is known as the *secondary elements*. More specifically, they describe the movement of satellites (*secondary elements*) around the earth (the *primary element* in this case).

It is useful to locate these orbits in relation to the movement of the earth around the sun. Seven parameters, often known as *Keplerian parameters*, can be used to characterize the movement of a satellite.

1.2.3. *Orbital parameters for earth satellites*

Let us take *(Oxyz)* as an orthonormal coordinate system, where *O* is the center of the earth, *Oz* its rotation axis and *(Oxy)* its equatorial plane, taking the direction of *(Ox)* as the intersection of the equatorial plane with the ecliptic plane (plane of the rotation of the earth around the sun).

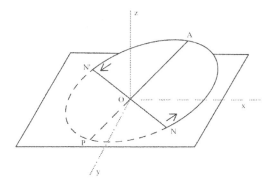

Figure 1.3. *The remarkable points of a satellite in orbit around the earth*

There are several remarkable points for a satellite in orbit (Figure 1.3):

– Apogee *A*, a point where the satellite is furthest from the earth;

– Perigee *P*, a point where the satellite is closest to the earth;

– the nodes N and N', the points where the satellite passes through the equatorial plane. In N, the satellite passes above the equatorial plane and in N', it passes below.

The movement of the satellite can be described with seven parameters.

Description of the orbit

The orbit is an ellipse given by:

– the semi major axis a, half the length of the major axis;

– eccentricity e: when $e = 0$, the orbit is described as circular and in the opposite case, it is described as elliptical.

Position of the orbit in relation to the earth

The parameters characterizing the position of the ellipse are:

– *inclination i* defined as the angle between the orbit plane and the equatorial plane, which is also the angle between the normal orbit and the rotation of the earth. By convention, inclination is between 0° and 90° when the satellite is turning in the same direction as the earth and between 90° and 180° when it is turning in the opposite direction;

– the perigee, which is the angle (ON, OP);

– the right ascension of the ascending node (Ω), which is simply the angle (Ox, ON).

Position of the satellite in the orbit

Now, when the orbit is defined, it is useful to specify where the satellite is located on it. To do so, it is necessary to specify an observation date and a place on the orbit:

– *the observation date t* is the moment the satellite is observed;

– *the average anomaly M* is the angle the satellite makes with the perigee. The angle is positioned according to the direction of the orbit (hence the subtle difference between N and N').

1.2.4. *Orbital perturbations*

The orbits described previously are perfect. In reality, two types of perturbations should be taken into consideration:

– *gravitational perturbations* only involve gravitational forces, but are caused by elements that are not taken into account by the previous model. Kepler's laws, in fact, consider that each planet has a point of volume equal to nil. In reality, the earth has a volume that is not insignificant, especially when low orbit satellites pass close by. The tide, the moon and the sun are also factors. Given that these forces derive from a potential, they do not reflect a satellite's loss or gain of energy. Nevertheless, the resulting deviation of the Keplerian satellite path is not always desirable;

– *non-gravitational perturbations* notably involve atmospheric drag and the pressure of (direct or indirect) solar radiation or radiation from another source (particularly in infrared).

1.2.5. *Maintaining and surviving an orbit*

The earth has large magnetic fields, which have an impact on its environment. Indeed, certain zones beyond the earth's surface emit radiation that is powerful enough to endanger the electronic components that pass through it.

Two energetic tori called the *Van Allen belts* circle the earth around the equator. These belts emit intense radiation that is very dangerous for the electronic equipment on board a satellite. The inner belt particularly affects altitudes between 1,500 and 5,000 km, while the outer belt affects those between 13,000 and 20,000 km.

Satellites orbit in four main regions:

– the LEO zone between the outer edge of the atmosphere and the inner Van Allen belt at an altitude of between 400 and 1,500 km;

– the *medium earth orbit* (MEO) zone between the two Van Allen belts at an altitude of between 5,000 and 13,000 km;

– the *high earth orbit* (HEO) zone whose apogee is beyond the Van Allen belts but whose elliptical orbit spans one or two of the previous zones;

– the *geostationary earth orbit* (GEO) zone, which can be considered as a specific case of HEO for geostationary satellites at an altitude of 35,786 km.

1.3. Time, time variation and coverage

Geostationary satellites are regularly criticized for their transmission time. Transmission time corresponds to the time it takes for electromagnetic waves to be propagated into space with a speed that is well estimated to be equal to that of the speed of light in space (300,000 km/s). It therefore takes 240 ms to go to a geostationary satellite and back. This explains why one of the main aims of low orbit satellites is to reduce the communication times of old systems. However, while geostationary satellite times correspond in an almost uniform manner to the some 35,800 km route separating them from earth, the picture become less uniform when the satellite views earth at a large angle, on the one hand, and when its movement is no longer synchronized with the rotation of the earth, on the other hand.

1.3.1. *Geometric data*

The closest point on the surface of the earth for radio communication with a satellite is located precisely at its vertical. Starting at this point on the surface of the earth and expanding out in concentric circles, reception gradually declines the greater distance from the satellite. In fact, given the lowest altitude at which satellites orbit (i.e. 200 km), the relief of our planet can be considered non-existent. The final coverage zone is therefore described as a disc on the surface of the earth; this disc is known as the *footprint* of the satellite [RES 95]. It is possible to give it a radius; however, a more important parameter is the radius of its solid angle of coverage, as illustrated in Figure 1.4. The solid angle is therefore the cone centered on the center of earth, from the axis passing through the satellite, whose angle of opening is θ.

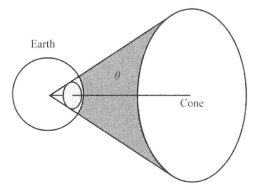

Dashed region: Intersection of the surface of the Earth and the cone

Figure 1.4. *Solid angle of coverage*

In fact, the coverage zone can be seen as the intersection of a circular cone starting from the center of the earth and from its surface. The knowledge of the angle θ is sufficient to be able to characterize it, if the vertical of the satellite, and thus the axis of the cone, is known. The advantages of this are twofold:

1) All users covered by a satellite with a coverage θ are at the intersection of the surface of the earth and the cone centered on earth with the angle θ whose axis passes through the satellite.

2) All satellites with a coverage θ covering a user are contained in the cone centered on the earth with angle θ whose axis passes through the user.

An essential element used to characterize coverage is the angle of elevation ω, which the user uses to view the satellite. In the event that the receiver is in the middle of an isolated area, and if the landscape is flat, they can pick up signals from the semi-sphere above their head. If plane P is tangential to the surface of the earth at the level of the receiver, the latter receives signals from the semi-space of the border P, which does not include the earth. However, the presence of the relief masks, in the majority of cases, at least one part of this space. In fact, a satellite situated too close to the horizon is changed by any obstacle that could get in its way: mountains, hills, buildings, trees, etc. To deal with this, the horizon line must be *raised* in some way so that the satellites are more *vertical*.

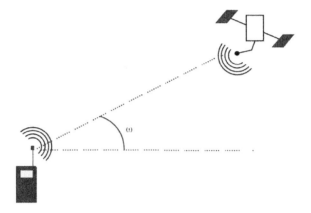

Figure 1.5. *Angle of elevation*

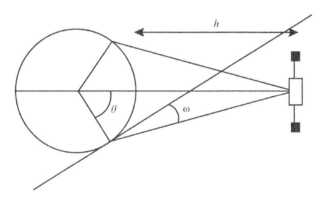

Figure 1.6. *The minimal angle ω*

Figure 1.6 shows how the minimal angle ω is measured. This angle is used to view a satellite.

If the altitude h of a satellite and the angle of elevation ω, which it is hoped is visible, is known, it is easy to determine the angle of coverage θ.

The various parameters are linked by the equation [ALT 99]:

$$h = R_T \left(\cos\theta(1 + \tan\theta \tan(\theta + \omega)) - 1\right) \quad [1.3]$$

where R_T represents the radius of the earth. This equation is also written as:

$$\frac{h + R_T}{R_T} = \frac{\cos\omega}{\cos(\theta + \omega)} \quad [1.4]$$

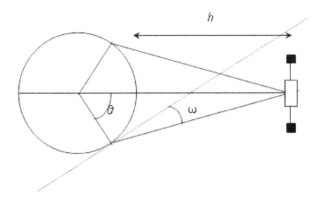

Figure 1.7. *Relation between θ, ω and h*

1.3.2. *Approximation of coverage*

The area covered by a satellite with a coverage θ has a surface area of $2\pi(1 - \cos\theta)R_T^2$ [ALT 99, CAP 03, SOR 99]. As the total surface area of the earth is $4\pi R_T^2$, at least $2/(1-\cos\theta)$ satellites are necessary to completely cover the earth.

It is, however, difficult to reach this level as each satellite covers a disc and the discs must overlap. Even if all satellites were immobile, finding the minimum number of discs required to cover a sphere is a mathematical problem that remains, to this day, unresolved [OCT 88]. More specifically, our problem [SOR 99], in particular when there are a large number of satellites, is that the cells covered by satellites are, in fact, hexagonal. Therefore, the area of a hexagon makes a ratio of about 0.827 with the area contained within the circumscribed circle. It is therefore necessary to add approximately 21% of satellites to the minimum number.

What is more, the coverage must take the movement of the satellite into account. So, when we want to create a permanent service, at least in a given zone, it must be guaranteed by either a geostationary satellite or have a family of satellites at its disposal: in this instance, we talk about *constellations* that ensure the global service. In this case, the movement of the satellites means that the covering satellite is constantly changing, even for an immobile user.

1.3.3. *Time interval between two successive intersatellite transfers*

A satellite transfer occurs when a user with an ongoing call switches from one satellite to another.

If a user comes close to the edge of a zone covered by a satellite, their coverage time by the latter will be very limited. This problem can be resolved by introducing larger recovery zones between neighboring satellites so that users at the outer edge of an area are covered by at least two satellites. The time interval between two intersatellite transfers is equal to the maximum time the same satellite can deal with a user.

1.3.4. *Time and time variation*

Links between users are often made with return channels. More specifically, in telephone communication, a conversation between an individual A and an individual B comprises two links $A-B$ and $B-A$. Therefore, if some time d_1 (respectively, d_2) is required for a user on earth A (respectively, B) to reach a satellite, the delay caused by the network and perceived by A while he or she asks a question to B is $2(d_1+d_2)$ (Figure 1.7).

In general terms, the additional perceived response time is two times the time required to cross the network. We can make a good approximation that this time is proportional to distance; radio waves move, including in the atmosphere, at a speed close to that of light in space. The transaction time via a geostationary satellite is therefore a minimum of $d = T+2(d_1+d_2)$:

$$\frac{4 \times 35\,800}{300\,000} = 0.477\,\text{s}$$

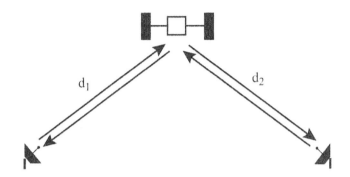

Figure 1.8. *Perceived response time (d) via a satellite with a processing time T*

The behavior of low orbit satellites is different from that of other categories of satellites: their times are very low and vary a great deal, unlike MEO and GEO satellites whose times are quite significant and vary very little. LEOs are therefore unquestionably able to break new boundaries with regards to applications. Nevertheless, some LEOs are not able to cope with the variations in perceived time, even if it means prematurely delaying the information received [WAL 84, BAL 80].

1.4. Orbital paths

The previous section described the criteria required for establishing a communication. These criteria are, however, local and cannot give a satisfactory perspective. So, in addition to establishing a communication, it is also necessary to guarantee a permanent service with a coverage that spans the whole planet [BAL 80]. There are therefore other equally important elements to be taken into account, including the specific restrictions relating to links with earth, networking and even cohabitation with concurrent systems. Some of these elements to be taken into consideration are:

– the *synchronization* of satellite paths with the rotation of the earth. For an observer on the earth, it is desirable for a satellite to stay in the same position (this is the case for geostationary satellites) or for it to periodically find itself at the same place over the course of the day;

– *earth-satellite transmission time* conditions the latency of the network (i.e. the duration of the transmission from start to finish);

– the *angle of penetration* of the earth's atmosphere that corresponds to the angle of elevation;

– the size of the *coverage cells*, the smaller the cells, the more important it is to reuse radio frequencies and the more the maximum number of users increases;

– the *total area covered by a satellite* (which can contain a number of cells), affecting the frequency of intersatellite transfers as well as the number of satellites required to cover a given zone;

– the *number of satellites required*, specifically to guarantee that the whole planet is covered for a service;

– the *required power of transmission*, the power can be used to increase the amount of information transmitted or to improve the portability of reception antenna (mobile telephones).

The characteristics of each of these systems will be analyzed in more detail.

1.4.1. *GEO-type systems*

These systems have a transmission time of 0.27 s, which is very high. They maintain a fixed position above the equator. Their high altitude enables them to cover a large area of the globe. Indeed, just three geostationary satellites are enough to cover almost the entire surface of the earth.

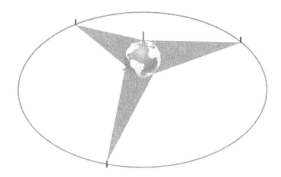

Figure 1.9. *GEO system, constellation with three satellites*

Nevertheless, these systems have significant coverage problems. They are unable, due to simple visibility problems, to cover the poles as well as any location with latitude over 81°. Furthermore, the angle of penetration in the atmosphere, in practice, makes it difficult to establish telecommunication links above 75°. However, even at latitudes between 45° and 75°, the angle of the satellite is small causing problems the moment the relief is not flat. High buildings in urban areas can therefore block coverage.

It would be possible to resolve this defect by using alternating and slightly inclined orbits. Satellites therefore complete one orbit in a figure of eight each day. These are non-geostationary geosynchronous paths.

1.4.2. *Elliptical systems*

In a certain sense, elliptical systems resolve the blocking problems GEOs have in high latitudes. Their orbit inclination of 63.4° enables them to position themselves above the equator. The angular velocity of the satellite on its elliptical orbit is reversely proportional to the square of its distance in relation to the center of the earth (the second of Kepler's laws). Therefore:

– *at its perigee*, the satellite moves very quickly in relation to the earth, even faster than a satellite with a circular orbit situated at the same location at the same time;

– *at its apogee*, the satellite moves very slowly in relation to the earth, slower indeed than a satellite with a circular orbit situated at the same location at the same time.

Satellites with an elliptical orbit are able to better center their footprint over the north and can make more precise adjustments (of coverage, synchronization). They do, however, regularly cross the *Van Allen* belts, considerably reducing their life expectancy.

However, both elliptical and geostationary systems are generally a very great distance from the earth, which has significant consequences on signal times and weakness. Transmission system designers have therefore looked for other solutions.

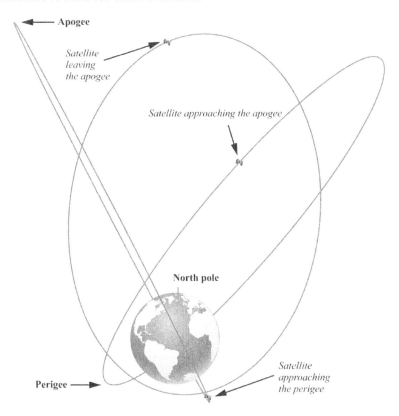

Figure 1.10. *Elliptical system, LOOPUS constellation*

1.4.3. *MEO-type systems*

MEO systems have a lower transmission time than their geostationary counterparts (110–130 ms at an altitude of about 13,000 km). Most MEO constellations have 10 or more satellites to guarantee a global coverage. Some systems only use an orbit with an inclination of zero, in which case coverage is excellent at the equator but rapidly declines as soon as the latitude increases. Other systems have slightly inclined orbits, requiring more satellites, but improving coverage.

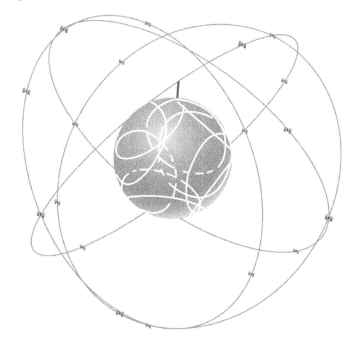

Figure 1.11. *MEO system, Spaceway NGSO constellation*

1.4.4. *LEO-type systems*

For LEO systems, transmission time varies between 20 and 25 ms; the precise time is at least as sensitive to the signals' angle of penetration in the atmosphere as to the precise altitude of the satellite. However, the devices are so close to the earth that a large number (at least 50) is required to cover the planet (due to the coverage reasons described in the previous section). The name *satellite constellation* mostly derives from this compulsory panoply. It is worth noting that to maintain coherent coverage zones, it is necessary that the movements of the satellites between themselves be synchronized very precisely.

Walker [WAL 71, WAL 84] defined two main types of constellations which continue to determine satellite constellation design today (Figure 1.12):

– Inclined constellations (Delta constellation by Walker) [WAL 71]

Inclined constellations have orbits with the same inclination i and with right ascensions of the ascending node Ω, which are regularly spaced out over a 360° area. In these constellations, both ascending and descending satellites guarantee that a zone is covered. Depending on the inclination i, their coverage will be better at the poles or at the equator;

Figure 1.12. *LEO system, Iridium constellation*

– Polar constellations (star constellation by Walker) [WAL 71, WAL 84, BAL 80]

Polar constellations have a series of orbits passing quite close to the poles and are organized in such a way that the ascending satellites (going from south to north) cover half the earth. The orbits all have the same inclination i that is close to 90° and have right ascensions of the ascending node Ω, which are regularly spaced out over a 180° area. Their coverage is therefore very dense at the pole and weaker at the equator. Additionally, it could be said that on the one side of the earth, all the satellites travel from South to North, whereas on the other side, they travel from North to South. As a result, satellites on the former orbit circle opposite and in reverse direction to those on the latter orbit. This phenomenon, which is known as *seams*, causes problems when communication between the satellites is sought.

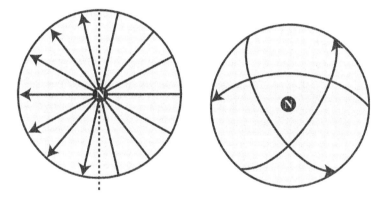

Figure 1.13. *Two types of popular constellations viewed from the pole*

1.5. Characteristics of cellular satellite systems

Cellular satellite systems are designed so that any authorized user, located anywhere on the earth, can have access and directly communicate with another user, either via a mobile phone or landline or via a public network

These systems are satellite constellations that are located on a number of interconnected orbital planes in a *ring* structure circling the earth. Each satellite generates several straight spotbeams that form cells on the surface of the earth. Unlike the cells in land networks that are fixed, these cells move with the satellite and sweep the surface of the earth.

Constellations of cellular satellite systems are designed in such a way that the satellite coverage footprints, which are symmetrically distributed, overlap to guarantee a continued coverage throughout the projected orbital path, as shown in Figure 1.14.

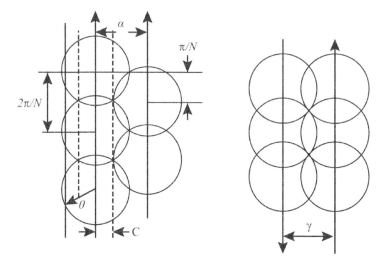

Figure 1.14. *Projection of coverage*

With half street width:

$$C_J = Cos^{-1}\left[\frac{Cos\,\theta}{Cos\,\pi/N}\right] \qquad [1.5]$$

where N is the number of satellites in the orbit, and θ is the radius of the satellite coverage.

The direction of this constellation is determined by the phase angle ψ separating the satellite planes and the angular separation α_a between the ascending nodes of the successive orbital planes. In every constellation, there are as many orbital planes as coverage interfaces (interfaces separating these planes) and as many satellites as coverage interfaces between the satellite planes.

If a constellation of a orbital plane a is designated for a $(a-1)$ corotational plane, the angle of separation γ between the last and the first planes in the contrarotational sector will be equal to $2C_j = 180 - (a-1)\,\alpha_a$ so that there will be neither gap nor radio silence at any point while the satellites of these planes are moving in opposite directions.

Likewise for the $(a-1)$ corotational planes, the angle of the interplane phase $\psi = \pi/N$, separating each satellite from its neighbors in the adjacent planes, will leave no radio silence or gap in the coverage interfaces between the planes.

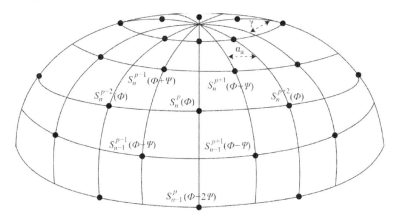

Figure 1.15. *Hemisphere of a constellation*

The common characteristics shared between satellite systems in order to ensure permanent mobile communication services are as follows [ALT 99]:

1) A "plurality" of corotational and contrarotational orbits that depend on the height of the orbit.

2) A number of equidistant satellites in each orbital plane.

3) The difference in the latitude phase between the satellites in the corotational sector is constant and depends on the height of the orbit.

4) The area of the coverage zone overlapped by each satellite with the neighboring satellites is a function of its latitude position.

The differences between the architectures of different satellite constellation systems in the world essentially reside in the inclination and the height of the orbits and, as a result, in the number of orbital planes and satellites in each plane.

1.6. The advantages of LEO systems

Three geostationary satellites located 36,000 km above earth guarantee a global coverage. Though they are perfectly suited to providing broadcasting services (radio, television), they are badly adapted to interactive multimedia services given that a transmission via a geostationary (GEO) satellite introduces a delay of about 270 ms thus making real-time applications difficult.

New low orbit (between 500 and 2,000 km of altitude) constellation systems avoid the ever-growing size of the frequency spectrum that is very sensitive on geostationary orbits. The times introduced by the propagation times, less than 20 ms, are compatible with communication protocols (TCP/IP, etc.), thereby ensuring the transparency of service in comparison to land networks. Moreover, the low altitude of the satellites makes the link budget more favorable and thereby reduces the emission power, the size of the satellites (LEO satellites generally have a mass of 500 kg in comparison to GEO satellites that weight several tons) and the size of the terminals (in particular, the antennae).

LEO constellations do, however, have some drawbacks. Countless satellites are required to guarantee a global coverage (the initial plan for the *Teledesic* system proposed a constellation comprising 840 satellites). LEO satellites, which are said to be scrolling as their rotation speed is higher than that of the earth, are only visible to a user on the earth for a few minutes, implying that the transmission regularly needs to be switched from one satellite to another to guarantee a continuity of service.

1.7. Handover in LEO satellite networks

Handover in LEO satellite networks has important characteristics including: a short propagation time, low power demands for the satellite or the user and a more effective allocation of the frequency spectrum. LEO satellites are considered to be the future of communication and information systems. However, LEO satellites are not stationary in relation to a fixed user on the surface of the earth. In fact, the speed of the satellite footprint on the earth is generally faster than the speed of the rotation of the planet and the user. The visibility period for a satellite in a cell is very brief and thus during one connection a user can be served by many spotbeams and satellites. Consequently, the link will switch from one spotbeam (potentially from a satellite) to another. This transfer is commonly known as handover [PAP 04a].

Consequently, maintaining a continuous communication on a LEO satellite system can require a change between one or several links as well as the Internet Protocol (IP) addresses of the *endpoints* of a communication [PAP 04a, CHO 00]. Handovers from the link layer and the network layer are therefore compulsory in satellite networks.

Managing the mobility of LEO satellites is therefore much more challenging than those of GEO or MEO systems. The mobility of a LEO satellite system is quite similar to that of cellular radio systems, though there are a number of differences. In fact, in the two systems, the relative position between the cells and mobile terminals is constantly changing, requiring the mobile terminals to switch (handover) between adjacent cells. The difference between the mobility of these two systems is as follows: in cellular systems, mobile terminals move through the cells, whereas in LEO systems, it is the cells that move through the mobile terminals. Cells in LEO satellite systems are larger than those in cellular radio systems. Moreover, the speed of a mobile terminal can be ignored in LEO satellite systems, since it is insignificant compared to the rotation speed of LEO satellites. However, unlike cellular systems on the earth where the movement of mobile devices is difficult to forecast, in LEO satellite systems, it is possible to forecast the movement of satellites and thus predicting the next satellite or spotbeam is relatively simple. At any moment, a real picture of the satellite constellation can be obtained, making it easier to select satellites in a communication path between two endpoints and therefore avoid unnecessary handovers. Handovers in satellite networks can be broadly categorized as presented in the following two sections.

1.7.1. *Link-layer handover*

Link-layer handover occurs when a change of one link or more is required between the communication endpoints because of the dynamic connectivity models of LEO satellites.

There are three types of link-layer handovers:

– *Spotbeam handover*: when users cross the border between the adjacent spotbeams of a satellite, an intrasatellite or spotbeam handover occurs. As the coverage zone of a spotbeam is relatively small, handovers between spotbeams are more frequent (every 1–2 min) [CHO 00];

– *Satellite handover*: when the connection established between a satellite and the attachment point of a user is transferred to another satellite, an intersatellite handover occurs;

– *Intersatellite links (ISLs) handover:* this type of handover occurs when the interplane of ISLs is extended due to temporary changes in the distance and the angle of visibility between the satellites in adjacent orbits. The progressive connections using these ISLs are derouted, thereby prompting an ISL handover.

The performance of the link-layer handover can be evaluated using two classic Quality of Service (QoS) criteria [TIA 01]:

– the probability of a blocked call (P_{b1}): the probability that a request for a new call be blocked;

– the probability of a forced termination (P_{b2}): the probability of the failure of the handover procedure causing the loss of an ongoing call.

There is a compromise between P_{b1} and P_{b2} in the different systems involved in the handover. Priority can be given by processing new calls and handover requests differently so as to reduce the ongoing calls being loss due to a failure in the handover procedure [OBR 99].

1.7.2. *Network-layer handover*

When one of the communication endpoints (satellite or user) changes IP address due to a change in the satellite coverage zone or the mobility of the user terminal, a network handover – or high-layer handover – is required to move the existing high-level protocols links (transmission control protocol (TCP), user datagram protocol (UDP), stream control transmission protocol (SCTP), etc.) to the new IP address. This is known as a network handover or high-layer handover [PUR 95]. Three different classifications can be used during the process of transferring a call [PAP 04a]:

– *hard-handover* systems: In these systems, the ongoing link is left before the next link is established;

– *soft-handover* systems: In these systems, the ongoing link is left only when the subsequent link is established;

– *signaling-diversity* systems: This system is similar to the *soft-handover* system, with the only exception that the signal flow of the old and the new links as well as user data are maintained by the old link during the handover procedure [PUR 95].

2

An Introduction to Teletraffic

2.1. Introduction

Communication networks (telephones, the Internet, local networks, etc.) have experienced exponential growth over recent decades. For operators, a vital issue is knowing how to control information flow in an optimal way in order to avoid congestion and offer users a good-quality, reliable and fast service. To design efficient procedures for controlling the circulation of information and correctly determine the software and hardware equipment necessary, in-depth knowledge about the properties of communication traffic is required.

The following discussion will describe the aspects of queuing theory that are fundamental for classic teletraffic theory [JAN 03, LI 07]. The majority of this information is dealt with in more detail in [COO 81, COO 90, SAA 61] and [WOL 89], which discuss teletraffic and queuing theories and teletraffic models developed since the work of Erlang.

We will start by describing the history of teletraffic and queuing theory, which are the foundations of the teletraffic technique. We will then explain the basic concepts of teletraffic theory: essentially birth–death processes and the Poisson process [RYB 05]. Finally, we will present some teletraffic models [STE 94].

2.2. The history of teletraffic theory and technique

The telephone was patented in 1876 and the first commercial telephone switchboard started operating in 1878 in New Haven, Connecticut. It consisted of a set of subscribers who could be connected two at a time. The need for teletraffic theory arose as soon as the number of subscribers increased to 3.

The first significant progress in teletraffic theory occurred in 1917 when A.K. Erlang, the Danish scientist, mathematician and engineer who worked for the Copenhagen Telephone Company, published an article that described a method and its use for determining a number of formulas that provided the basis for a modern teletraffic theory and technique.

Later, when operations research was invented during World War II, Erlang's methods and models were incorporated into queuing theory, and these two topics (queuing theory and teletraffic theory) are now closely linked. A great deal of attention has been paid to these theories in the literature.

2.2.1. *Queuing theory*

Queuing theory is a mathematical systems theory that provides clients with arrival times and random demand services. If servers are not available during client arrivals, then a service queue is formed, giving the theory its name.

This theory dates back to 1909 when A.K. Erlang laid out the foundations of his research into telephone traffic. His research was then integrated into operations research. Unfortunately, publications about queuing theory started to use increasingly mathematical language, resulting in the theory no longer being used. The situation has since changed: queuing theory has been applied to performance evaluation. For this application, it appeared that even relatively simple queuing models provided results that closely corresponded to real observations. A rapid evolution in queuing theory followed and the theory was subsequently applied to evaluating the performance of information and communication systems. Due to various pieces of research, queuing theory is widely used today and has countless applications.

The traditional theory is based on the Markov chain, that is to say that all events (such as arrival, end of service and a change in queuing) depend exclusively on the current state of the system and not on its previous behavior. This simplifies not only the mathematical processing, but also data collection because only averages are required, such as the average service time, the average arrival per unit of time and so on [STE 94, HEY 82, BEY 14, BAY 14].

2.2.2. *Teletraffic theory*

The mathematical analysis of traffic in communication networks is an old discipline dating back to research carried out by the Danish engineer A.K. Erlang in 1917. This approach, and other approaches by different researchers, provided the main mathematical tool of dimensioning, which was used by network operators and builders until about the 1990s [COO 98].

The mathematical approach explored by Erlang and other researchers and engineers after him was essentially Markovian. This means that the approach describes traffic using a simple random process model, Markov chains, which has an advanced and strong mathematical theory (Andreï Markov, 1856–1922, was a Russian mathematician who made important contributions to probability theory).

In simple terms, a Markov chain is a series of random events in which the probability of a given event is exclusively dependent on the events that immediately precede it. In the context of communication networks, Erlang's Markovian approach implies that the statistical laws characterizing the traffic are Poisson laws: the Poisson law is one of the most popular and simple probability or statistical laws (its name derives from the French mathematician Denis Poisson, 1781–1840). Poisson hypothesis proved to work for telephone traffic (the random events are subscribers' calls which occur randomly and whose duration is also random).

This type of traffic modeling enabled adapted control procedures to be put in place. Until recently, communication network control involved admission control, when the operator refuses a user access to the network when the latter cannot guarantee a predefined quality of service.

2.3. Basic concepts

Let us consider a basic model to be a system where the calls arrive at random moments and each call requires the use of a channel. If a channel is available, the call holds it for a random period of time, known as the *holding time*, but if no channel is available the blocked call takes some specific actions that include blocking, trying to connect the call again or putting the call in a queue (using queuing theory terminology, the calls are the clients, the channels are the servers and the holding time is the service time). The aim of teletraffic theory is to determine a good description of the random teletraffic (a description of the statistical or stochastic properties, call arrival times and holding times) and to create formulas describing the performance of the system (such as the probability of blocking, the fraction of lost calls and the average queuing time) in accordance with the demand and the number of channels. This theory is then adapted and applied to designing and managing real telecommunication systems: this is the teletraffic technique.

The essential concept behind the teletraffic technique is that it is stochastic; the mathematics used henceforth will thus be probabilities, statistics and stochastic processes. These mathematical processes will now be summarized as briefly as possible before they are applied to develop basic teletraffic theory.

2.3.1. *The birth–death process*

To understand these concepts, let us consider the following model: calls arrive according to a stochastic process to a group of S identical channels. If an incoming call finds an available channel, it holds it for a random holding time, after which the call liberates the channel and it becomes available for another user. If all channels are held, the incoming call is rejected and, in this case, takes one of the specific actions described below.

Let us suppose that P_j denotes the probability that the system is at state j; therefore, P_j is the probability that the number of calls present (in service or, if the model allows it, waiting in a queue for a channel to become available) is equal to j. Let us suppose that while the system is at state j, the call arrival rate is λ_j, and the call departure rate is μ_j. Therefore, it is possible to show that, under certain conditions that must meet arrival and departure processes, the following equations determine the probability of the states according to the rates λ_j, and μ_j:

$$\lambda_j P_j = \mu_{j+1} P_{j+1} \qquad (j = 0, 1, 2, \ldots) \tag{2.1}$$

and

$$P_0 + P_1 + \ldots = 1 \tag{2.2}$$

A successive solution to equation [2.1] for each P_j gives us:

$$P_j = \frac{\lambda_0 \lambda_1 \ldots \lambda_{j-1}}{\mu_1 \mu_2 \ldots \mu_j} P_0 \tag{2.3}$$

P_0 is calculated from the normalization equation [2.2]:

$$P_0 = \frac{1}{1 + \dfrac{\lambda_0}{\mu_1} + \dfrac{\lambda_0 \lambda_1}{\mu_0 \mu_1} + \ldots} \tag{2.4}$$

A stochastic process described by equation [2.1] is called the "birth–death process". The probabilities P_j described by [2.1] are the probabilities of average times; P_j is the amount of time taken by the system to pass to state j. Other interesting probabilities are the probabilities of the average number of subscribers A_j, the number of subscribers arriving to the system while it is at state j. In general, these two probabilities are not equal, but when clients arrive according to a Poisson process:

$$A_j = P_j \quad (j = 0, 1, ...) \qquad [2.5]$$

This equality reflects the *Poisson Arrivals See Time Averages* (PASTA) theorem (see [COO 98]).

2.3.2. *Poisson process*

A common assumption in teletraffic theory is that call arrivals follow a Poisson process.

Let us suppose that time is divided into intervals of equal size Δt and that:

1) there will be at most one arrival in each interval;

2) the probability of an arrival at a given interval is proportional to Δt;

3) intervals are statistically independent of each other.

Let us suppose that the random variable x is the period of time spanning from the instant t_0 (time 0 sec) to the instant the next call arrives. Let us calculate the probability $P(x > t)$ that no call arrives during the interval $[0, t]$. Let us imagine that the interval $[0, t]$ is divided into n intervals, each of a size $\Delta t = t / n$. If the proportionality constant is λ and is supposed to be 2, the probability that an arrival occurs in any interval is Δt is $\lambda \Delta t = \lambda t / n$, and from point 3, we can say that the probability that no arrival will occur in the n intervals of

[0, t] is $(1-\lambda t/n)^n$. Let us now pass from discrete time to continuous time $\Delta t \to 0$, and therefore $n \to \infty$, thereby giving:

$$P(x > t) = \lim_{n \to \infty} \left(1 - \frac{\lambda t}{n}\right)^n \qquad [2.6]$$

This is a well-known limit in mathematics, which is equal to $e^{-\lambda t}$. Let us designate the distribution function x with $F_x(t) = P(x \leq t)$; therefore, from [2.6]:

$$F_x(t) = 1 - e^{-\lambda t} \qquad t \geq 0 \qquad [2.7]$$

A random variable with a distribution function given by equation [2.7] is said to be exponentially distributed from the average value $E(x) = 1/\lambda$ and the process describing the arrivals with independent and identically distributed (i.i.d.) intervals of time, whose distribution function is given by [2.7], is called a Poisson process. The average value $E(x)$ is the supposed proportionality constant.

An important property of all exponentially distributed random variable is the Markov property (memoryless), expressed by Stewart [STE 94]:

$$P(x > y + t \mid x > y) = P(x > t) \qquad [2.8]$$

In the context of the birth–death process described by equation [2.1], if call arrivals follow a Poisson process with a rate λ, then the instantaneous birthrate λ_j while the system is at state j is the same for all states; therefore, $\lambda_j = \lambda$. Similarly, the instantaneous death rate μ_j is found by supposing that the call durations (the periods a channel is held) are i.i.d. exponential random variables with an average value $\tau = 1/\mu$ and by applying the exponential distribution property that states that the minimum of a group of independent and exponentially distributed variables is also exponentially distributed with a rate equal to the sum of the original rates, which gives $\mu_j = j\,\mu$ when there are, simultaneously, j ongoing calls [HAL 83, ROS 93, ROS 83].

2.4. Erlang-B and Erlang-C models

When we have a Poisson arrival process with a rate λ and the holding times are exponential with an average $1/\mu$, then the probabilities of the states P_j are determined by equation [2.1] with $\lambda_j = \lambda$ and $\mu_j = j\mu$ when $j \leq S$ and $\mu_j = S\mu$ for $j > S$. The solution of equation [2.3] for $j \leq S$:

$$P_j = \frac{(\lambda/\mu)^j}{j!} P_0 \qquad (j = 0,1,...,S) \qquad [2.9]$$

2.4.1. *Blocking probability and the Erlang-B formula*

In this section, we will look at a system that has S channels of communication. If the S channels are held, incoming calls are lost (no tone or holding tone, for example). This is known as a *blocked calls cleared* (BCC) system. We are going to calculate this blocking probability according to the number of channels available and the traffic. New calls arrive following a Poisson process.

Therefore, in this case, $P_j = 0$ for all $j > S$, and therefore equation [2.2] gives:

$$P_0 = \frac{1}{\sum_{k=0}^{S} \frac{(\lambda/\mu)^k}{k!}} \qquad [2.10]$$

By putting $a = \lambda/\mu$, which represents the offered traffic, we obtain:

$$P_j = \frac{\frac{a^j}{j!}}{\sum_{k=0}^{S} \frac{a^k}{k!}} \qquad (j = 0,1,...,S) \qquad [2.11]$$

The group of probabilities defined by equation [2.11] is called the *Erlang loss distribution* [ROS 95] and was demonstrated by Erlang in 1917.

In particular, the blocking probability of a system with S channels and for traffic a is therefore written as $B(S, a)$. It is equal to the probability of finding itself in the state S, $B(S, a) \equiv P_S$ and is obtained by the following equation:

$$B(S,a) = \frac{\dfrac{a^S}{S!}}{\displaystyle\sum_{k=0}^{S} \dfrac{a^k}{k!}} \qquad [2.12]$$

This formula is very important in telecommunications and is called the Erlang-B formula.

In this model, the traffic flow is given by:

$$a'_{BCC} = \sum_{j \leq S} j P_j \qquad [2.13]$$

By substituting equation [2.11] into equation [2.13], we can easily obtain:

$$a'_{BCC} = a[1 - B(S,a)] \qquad [2.14]$$

This enables us to say that the traffic flow equals the product of the offered traffic and the fraction of the latter that is not lost, and therefore:

$$a'_{BCC} = a - aB(S,a) \qquad [2.15]$$

It is difficult to calculate the numeric values for Erlang-B directly from equation [2.12] for large values of S and a. However, it is easy to demonstrate (see Appendix 1) that:

$$B(n,a) = \frac{aB(n-1,a)}{n + aB(n-1,a)} \qquad (n = 1,2,...,S; B(0,a) = 1) \qquad [2.16]$$

and to use it to write a program. This algorithm is stable and fast.

2.4.2. Queuing probability and the Erlang-C formula

If we consider a system in which blocked calls can be put into a queue before being served, the queuing probability can be defined. It is also supposed in this case that new incoming calls follow a Poisson process.

With this system, we still have:

$$\lambda_j = \lambda \qquad [2.17]$$

In contrast, for the probability of the call ending we have:

$$\mu_j = \begin{cases} j\mu & (j \leq S) \\ S\mu & (j > S) \end{cases} \qquad [2.18]$$

By replacing λ_j and μ_j in equation [2.3], we obtain:

$$P_j = \begin{cases} \dfrac{a^j}{j!} P_0 & (j=1,2,\ldots,S-1) \\ \dfrac{a^S a^{j-S}}{S! S^{j-S}} P_0 & (j=S,S+1,\ldots) \end{cases} \qquad [2.19]$$

P_0 is given by:

$$P_0 = \dfrac{1}{\sum_{k=0}^{S-1} \dfrac{a^k}{k!} + \sum_{k=S}^{\infty} \dfrac{a^S}{S!} \dfrac{a^{k-S}}{S^{k-S}}}$$

$$P_0 = \dfrac{1}{\sum_{k=0}^{S-1} \dfrac{a^k}{k!} + \dfrac{a^S}{S!} \sum_{k=S}^{\infty} \dfrac{a^{k-S}}{S^{k-S}}}$$

We can write:

$$P_0 = \dfrac{1}{\sum_{k=0}^{S-1} \dfrac{a^k}{k!} + \dfrac{a^S}{S!} \sum_{k=0}^{\infty} \left(\dfrac{a}{S}\right)^k}$$

We obtain for $a/S < 1$:

$$P_0 = \frac{1}{\sum_{k=0}^{S-1} \frac{a^k}{k!} + \frac{a^S}{S!(1-a/S)}} \qquad (a<S) \qquad [2.20]$$

Therefore, in this case, the probability of a blocked call, which is also the probability of queuing, is equal to $P_S + P_{S+1} + ... \equiv C(S, a)$:

$$C(S,a) = \frac{\frac{a^S}{S!(1-a/S)}}{\sum_{k=0}^{S-1} \frac{a^k}{k!} + \frac{a^S}{S!(1-a/S)}} \qquad (a<S) \qquad [2.21]$$

This formula is also very important and is called the *Erlang-C* formula.

In this case, the traffic flow is equal to:

$$a'_{BCD} = \begin{cases} a & (a<S) \\ S & (a \geq S) \end{cases} \qquad [2.22]$$

In the same way as the Erlang-B formula, it is possible to give another formula for this case. It is easy to demonstrate (see Appendix 1) that:

$$C(S,a) = \frac{SB(S,a)}{S - a[1 - B(S,a)]} \qquad (a<S) \qquad [2.23]$$

3

Channel Allocation Strategies and the Mobility Model

3.1. Introduction

When designing cellular satellite systems it is very important to establish the best message routing technique that maintains a reasonable level of traffic quality. The constellation geometry of these networks is dynamic and has short links as the satellites move quickly in comparison to stations on the earth. Several strategies for managing radio resources have been developed to better meet the requirements of users of mobile communication networks, particularly channel allocation [LEE 98, ROS 95].

In fact, extensive research has been published in the literature about channel allocation in cellular telephone systems, bearing witness to the fact that the frequency spectrum available is congested. The most widespread techniques are fixed channel allocation (FCA) and dynamic channel allocation (DCA).

Another extremely important problem with regard to managing resources in mobile telephone systems is spotbeam handover. This very frequent problem has been the subject of a number of pieces of research which have put forward different strategies and methods for managing handover requests.

This chapter will first present the foundations of channel allocation. We will then describe the strategies that guarantee priority to handover requests.

Whether a study is theoretic or if it simulates the impact of channel allocation or a handover request management strategy, it requires the mobility model of the network being researched to be defined. The mobility model adopted in our research will, therefore, be described in the penultimate section of this chapter and then analyzed in the final section.

3.2. Channel allocation techniques

A number of channel allocation techniques have been put forward in the literature. With regard to the problem of frequency assignment, *Koster* presents in [KOS 99] a detailed and very interesting study into algorithms and mathematical models. In reference [TEK 91], *Tekinay* and *Jabbari* present the three main categories of these techniques, which are:

– FCA;

– DCA;

– hybrid channel allocation.

As its name suggests, the latter technique is a hybrid of the other two – fixed and dynamic – which are the most widely used basic channel allocation techniques in mobile communication networks.

These techniques must meet the frequency reuse criterion: two different cells can use one channel, provided that they are a reasonable distance from each other (called the reuse distance), enabling tolerable levels of interference [SAH 98].

A brief description of these two strategies is given below.

3.2.1. *Fixed channel allocation techniques*

The FCA technique [TEK 92, SAN 95] permanently assigns a set of channels to each cell, respecting the permitted reuse distance. A call can only be served by an available channel belonging to this set. If a new call started in a cell finds that none of its nominal channels are free, the call is blocked and lost.

For uniform traffic conditions, the whole set of M channels is divided into equal groups, each composed of S channels [SAH 98].

$$S = \frac{M}{K} \quad \text{with} \quad K = \frac{D^2}{GR^2} \qquad [3.1]$$

In [3.1], parameter K is the number of cells forming the FCA cellular pattern [BAY 14], D is the frequency reuse distance and R is the side of the cellular hexagon. The space of this group is repeated in a mosaic-like manner, guaranteeing that the region is covered.

Henceforth, $F_D(x)$ will designate the group of channels assigned to a cell x according to the FCA technique.

Using FCA for non-uniform traffic means that planning the network is complex, as more capacity has to be assigned to cells where we expect there to be more traffic. For low earth orbit (LEO) satellite communication networks, there is no use for this sort of planning as the traffic offered to a given cell is unpredictable due to the fast movement of the satellite in relation to the earth. For this reason, the DCA approach is much better suited to LEO mobile satellite system (MSS) systems; this approach will be described below.

3.2.2. *Dynamic channel allocation techniques*

A large number of DCA techniques have been developed in the literature [AKY 99, TEK 92, DEL 99, EFT 98]. We will focus particularly on dynamic allocation presented by Del Re et al. [DEL 95, DEL 99].

DCA automatically allocates the channels of Hertzian transmissions, thereby guaranteeing the lowest levels of interference between the cells in the network [EFT 98].

With this strategy every system channel can be temporarily assigned to any cell, provided that the reuse distance constraint is met.

Let x be the arrival cell of a new call, $I(x)$ is the set of interference cells of x (i.e. the cells at a distance less than D to x) and $\Pi(x)$ is the set of channels available in x (i.e. the channels that are not used in x nor in cells belonging to $I(x)$).

The DCA technique selects the channel, which must be assigned to a new call arriving in the cell, according to the *minimum cost* criterion [WAN 93, DEL 94, DEL 95, DEL 96, DEL 97, BOG 01]:

$$C_x(i^*) = \min_{i \in \Pi(x)} \{C_x(i)\} \qquad [3.2]$$

The *cost function* $C_x(i)$ is defined as follows:

$$C_x(i) = q_x(i) + \sum_{k \in I(x)} \{C_x(k,i)\}, \quad \forall i \in \Pi(x) \qquad [3.3]$$

where the contribution of the channel cost $i \in \Pi(x)$ due to the interference cell $k \in I(x)$, $C_x(k,i)$ is given by:

$$C_x(k,i) = u_k(i) + 2(1 - q_k(i)) \qquad \forall k \in I(x) \qquad [3.4]$$

With:

$$u_K(i) = \begin{cases} 1 & \text{if } i \in \Pi(k) \\ 0 & \text{otherwise} \end{cases} \qquad [3.5]$$

It is worth pointing out that $C_x(k,i)$ only has four values:

$$C_x(k,i) = \begin{cases} 0 & \text{if } i \notin \Pi(k) \text{ and } i \notin F_D(k) \\ 1 & \text{if } i \in \Pi(k) \text{ and } i \notin F_D(k) \\ 2 & \text{if } i \notin \Pi(k) \text{ and } i \in F_D(k) \\ 3 & \text{if } i \in \Pi(k) \text{ and } i \in F_D(k) \end{cases} \qquad [3.6]$$

The first term in $C_x(k,i)$ takes the availability of the channel in the cell k into account, while the second term takes the channel's belonging to the nominal group of the cell k (i.e. $F_D(k)$) into account. This second term is reinforced by a factor 2 so that the FCA distribution can be followed as much as possible.

Similarly, the term $q_x(i)$ is introduced into formula [3.3] so that channels belonging to $F_D(x)$ are favored for selection as far as possible, i.e. the set of channels attributed to x according to FCA. In other words, DCA strategy chooses the channel that becomes locked in the minimum number of interference cells in $\Pi(x)$. If $\Pi(x)$ is empty, the call is blocked.

To improve the performance of DCA, each time a call ends in a cell (due to either the call ending physically or a failure in the handover procedure), a channel is freed according to a no assignment criterion [DEL 95, DEL 96] with a cost function complementary to that used in the allocation phase. To be freed in x, it chooses the channel that becomes available in the greatest number of interference cells and – as far as possible – a channel that does not belong to the FCA pattern of x (see Appendix 2). If the channel chosen is different to the channel where the call really ended, the ongoing call in the first channel must be transferred to the latter.

3.3. Spotbeam handover and priority strategies

3.3.1. *Spotbeam handover*

Given that satellites move faster than the surface of the earth, LEO telecommunication satellite networks are blighted by the extremely important problem of handover: mobile users (MUs) with ongoing calls can switch spotbeams and potentially even satellites. In this situation, a new channel must be automatically assigned to MUs in the destination spotbeams so that they can have a conversation without the connection cutting out. If no channel is available in the transit cell, the call is lost.

In non-geostationary satellite systems, the handover procedure rate is conditioned by the speed of the satellite (and thus by the altitude of the satellites in the constellation). In LEO satellite networks; this procedure occurs very frequently for both fixed and MUs. The handover procedure fails when no channel is free in the transit cell. This failure brings ongoing calls to a forced end; users are frustrated by this cut in the service and their demand for *Quality of Service* (QoS) is infringed [ITU 95]. Therefore, one of the principle objectives when designing a communication network must be a low probability of established calls cutting out.

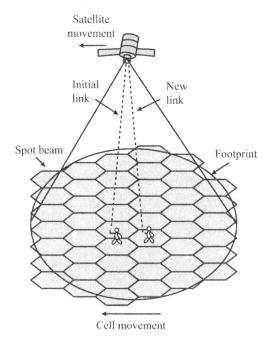

Figure 3.1. *Cellular handover*

To improve the cellular QoS , a number of methods giving priority to handover requests over new call requests have been proposed to reduce the probability of handover failure. This occurs at the expense of a sometimes tolerable rise in the probability of new calls being blocked [YAC 02].

3.3.2. *Priority strategies for handover requests*

In LEO satellite networks, spotbeam handovers frequently occur due to the relatively small size of the Hertzian spotbeam zone and the relatively high speed of the satellites [SAH 98]. These frequent handover requests could cause a call to be blocked if no resources are available in the destination spotbeam. From the user perspective, an ongoing call being blocked is generally seen to be less desirable than a new call being blocked. Priority of service can, therefore, be given to handover requests [TEK 92, KÜP 05].

A number of strategies (presented in the literature) for managing handover requests guarantee the success of handover procedures so that ongoing calls are not lost during the transition from one spotbeam to another [DEL 93, PAL 95, CHE 96, AKY 99, GUR 91]. Other strategies try to ensure a good QoS by giving priority of service to handover requests over new calls, without nevertheless guaranteeing that the procedure will succeed [LIL 05, GUE 87].

These priority systems can be broadly divided into two groups according to the two following concepts:

3.3.2.1. *Concepts of guaranteed handover*

In [MAR 91], *Maral et al.* present a guaranteed handover (GH) system in which each call tries to reserve a channel in the transit cell next to the one it is in. If no channel is free, the request is put in a queue. Therefore, a new call arriving in the cell i tries to reserve a channel not only in the cell i but also in another channel in the transit cell $i+1$. This strategy results in a considerable fall in the probability of ongoing calls being blocked, but at the expense of the probability of new calls being rejected, which reaches values that are too high. To improve the allocation of resources, a number of studies have proposed modified GH techniques: *elastic channel locking* (ECL) [XU 00], *time-based channel reservation algorithm* (TCRA) [BOU 03, BOU 01, BOU 03] and *dynamic Doppler-based handover prioritization* (DDBHP) [PAP 03, PAP 04].

3.3.2.2. *Concepts of guaranteed priority to handover*

Systems guaranteeing priority to handover requests can be divided into four groups according to the concept they adopt:

– handover with guard channel: the system reserves a number of channels exclusively for handover requests;

– handover with queuing: the system uses the benefits of the overlap zone between cells – the MU can be served by two cells in this zone [DEL 99] – to put, for a certain period, handover requests in a queue for a free channel in the transit cell;

– handover based on channel rearrangement: used exclusively with DCA [DEL 99], at the end of each call the system rearranges the channels to free the channel that will become available in the maximum number of adjacent cells;

– handover with guard channels and queuing: combines the advantages of using guard channels and handover queues.

We will now give a brief description of the first two categories:

– *The concept of handover with guard channels*

Introduced in the 1980s [BEL 70], this concept offers a generic system that improves the probability of handover requests succeeding by reserving a number of channels exclusively for them. The other channels will be used for both handover requests and new calls. The major disadvantage of this method is that attributing guard channels means there are fewer channels available for new calls, therefore, there is a significant rise in the probability of new calls being rejected and thus a perceptible reduction in the total traffic capacity. This drawback can be avoided by enabling new calls to be put in a queue. Intuitively, it can be said that this method is feasible since new calls are considerably less sensitive to times than handover requests.

Another problem with using guard channels, particularly with FCA strategies, is the risk that the allocated spectrum will be inefficiently used. A careful estimation of the optimum number of guard channels is, therefore, required. In flexible or DCA strategies, the concept of guard channels is presented in a different way: the number is dynamically determined according to traffic capacity. Let us cite, for example, *dynamic channel reservation* (DCR) introduced by *Wang et al.* [WAN 01], in which the reservation of channels in a cell is determined dynamically by constantly estimating the probable number of handover requests the calls in the previous cell can initialize. This number is calculated using the position of users and their handover probabilities (the probability that users with ongoing calls will need a handover procedure). We have adopted this strategy in our study and it will be explained in more detail in the next chapter.

– *Organizing handover requests into queues*

This strategy takes advantage of the overlap zone between cells, a zone where calls can be served by two stations at the same time. The coverage zones of two adjacent cells overlap, and there is therefore a common zone known as the overlap zone. This zone gives handover requests that cannot find a free channel in the transit zone an additional waiting time to be served, rather than the call coming to an abrupt end at the limit of the current cell. If in the meantime a channel becomes free in the transit cell, the call is transferred to the latter. However, if the MU reaches the limit of the overlap zone without being served, the call is lost.

Calls are put in a queue in a specific order. A number of *queuing strategies* have been presented in the literature. The most popular strategy is first in first out (FIFO) [AKY 99, DEL 95, HON 86]: it ranks calls in order of their arrival times. The call spending the most time waiting for a channel to be freed in the transit cell is placed at the top of the queue. The more complicated *measurement-based priority scheme* (MBPS) strategy was put forward by *Tekinay et al.* [TEK 92, TEK 91]. This method uses a *non-preventative* dynamic priority technique, in which the priority of handover requests is determined by the level of power the satellite receives from the calls in question (queuing) from their current spotbeams. The *last useful instant* (LUI) [KIA 08] strategy is considered ideal. It is based on the maximum queuing time. We have adopted this last strategy in our study and it will be explained in more detail in the next chapter.

3.4. Mobility model

To study the process of generating handover requests toward a cell and to evaluate the impact of radio resource management strategies on network performance, it is necessary to define a user mobility model.

Mobility models have, therefore, been the subject of a number of studies for both land and satellite networks. For example, in [QUI 04] Quiles presents a mobility model for land networks. His research is very interesting as it includes the possibility that users with ongoing calls stop for a certain amount of time during the communication. For satellite networks, which have a very different model to land networks, a model presented in [MAR 98, COO 90, KOS 99] has been widely used in the literature. Its major disadvantage is that it assumes that calls are generated exclusively in the central zone of cells. In 1999, Del Re *et al.* presented a more realistic mobility model which did not fail to take into account the *junction* of network cells [DEL 99]. This is the model which we selected for our research and will describe it below.

To model the mobility of users in a communication network, the following aspects must be taken into account: (1) the propagation conditions of the satellite's radio channels; (2) the movement of the user in relation to the cells; and (3) the geometry of the cells and their arrangement, that is the network topology.

To reduce the analytical complexity of the model, an approach that is commonly used in the literature is to ignore the propagation aspects [ZHA 95, WAL 84, MCN 04, DAS 02]. Consequently, this study will only take into account the movement of the user and the network topology. The traffic considered is telephonic.

The coverage zone has been divided into several cells and each cell is sent by a satellite antenna. We considered that the cells are hexagonal and regular (see Figure 3.2).

In practical applications, the side of the cell R varies between a minimal value R_{min} (when the satellite is at the zenith) and a maximal value R_{max} (corresponding to the minimum angle of elevation with which the MU can see the satellite), following the position of the satellite in relation to the MU.

Let us suppose that R is equal to $0.5(R_{min} + R_{max})$. This supposition will enable us to obtain a simple analytic characterization of user mobility.

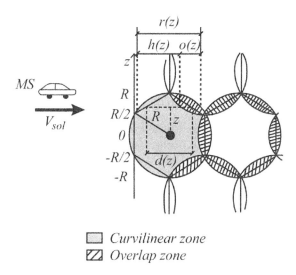

☐ Curvilinear zone
▧ Overlap zone

Figure 3.2. *Geometry of cellular networks with hexagonal patterns*

The cell where the call starts is called the source cell; any consecutive cell reached by an MU with an ongoing is called the transit cell.

Let us take a given cell x, the index $I = 1$ will be used for statistical parameters related to the calls initialized in this cell, while the index $i = 2$ will be used for parameters related to the calls transferred to x.

The relative satellite-MU movement can only be approximated by the *satellite ground-track* (i.e. the vector V_{sat}), due to its high value in relation to the other speed components (MU movement and earth rotation). Consequently, the relative satellite-MU movement has a fixed direction in relation to the satellite ground-track cast by satellites (see Figure 3.2).

MUs (and the calls they generate) are considered uniformly distributed over the simulation zone [MAR 91, HON 86, MCN 04, DAS 02, KIA 11a]. Consequently, the probability of a new call arriving is the same at each point on the cellular satellite network.

According to the aforementioned suppositions:

– When a new call arrives, every cell in the system has the same probability of being the source cell of this call.

– Once the source cell of a call is defined, it is given a random position z. The *probability density function (pdf) f(z)* of z is obtained, taking into consideration that the active MUs are uniformly generated in a cell and that they move in straight lines.

– A new call in its source cell belongs to an elementary horizontal beam with side $d(z)$ and height dz (see Figure 3.3) according to a probability given by the relation between the surface of the beam [$= d(z)dz$] and the surface of the cell ($= 3\sqrt{3}R^2/2$)

Therefore, $f(z)$ is given by:

$$f(z) = \frac{d(z)}{3\frac{\sqrt{3}}{2}R^2} \qquad [3.7]$$

where *d(z)* is equal to:

$$d(z) = \begin{cases} \sqrt{3}R & \text{if } |z| \leq \frac{R}{2} \\ 2\sqrt{3}(R-|z|) & \text{if } R \geq |z| > \frac{R}{2} \end{cases} \quad [3.8]$$

– Once the offset z of the MU is chosen in the source cell, the distance covered in this cell by the MU from the moment the call arrives is uniformly distributed between zero and $d(z)$.

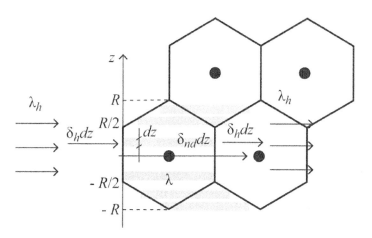

Figure 3.3. *Flow conservation of requests entering and leaving a cell*

In accordance with the uniform spatial generation for new call requests, the offset of an MU in a transit cell is uniformly distributed. This implies that the *pdf* of the offset z according to which an MU travels across the source cell or a transit cell can be given by:

$$f_i(z) = \begin{cases} f(z) & \text{if } i = 1 \\ \dfrac{u[z+R] - u[z-R]}{2R} & \text{if } i = 2 \end{cases} \quad [3.9]$$

With

$$u(x) = \begin{cases} 1, & x \geq 0 \\ 0, & otherwise \end{cases}$$

We could summarize the proposed LEO mobility model as follows:

1) MUs travel across a cellular network with a relative speed (i.e. vector V_{sat}), arranged in relation to the cellular plane as shown in Figure 3.2 (repeated below).

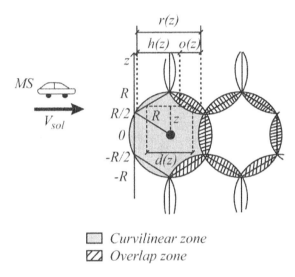

☐ Curvilinear zone
▨ Overlap zone

2) When a handover occurs, the neighboring cell in the direction of the relative satellite-MU travel will be the transit cell.

3) The MU travels across the cellular network with an offset that is uniformly distributed across the network.

4) From the moment a call arrives in a cell where z is the offset of the MU, the latter travels a distance:

i) uniformly distributed between zero and $d(z)$ if the cell is the source cell of the call;

ii) exactly equal to $d(z)$ if the cell is a transit cell.

To characterize the (relative) mobility of the user in LEO mobile satellite networks, the parameter α is defined as:

$$\alpha = \frac{\sqrt{3}R}{V_{sat}T_m} \qquad [3.10]$$

where T_m is the average duration of the call.

The parameters R and V_{sat} depend on the altitude of the satellite constellation; which is more, R also depends on the *half-power beam width* of the spotbeams of satellite antennae. The typical values of α are 0.20–0.60 for LEO MSS, for $T_m = 3$ min.

The smaller the α is, the more frequent the handover requests are throughout the duration of the call. Specifically, in the IRIDIUM case mobility considered in this study ($R = 212.5$ km, $V_{sat} = 26\,600$ km/h) α is roughly equal to 0.27, for $T_m = 3$ min.

3.5. Analysis of the mobility model

Let us continue to consider a given cell x. We denote by t_{mc2} the time required for an MU to cross from one end of the cell to the other (i.e. the amount of time the mobile stays in a cell) and by t_{mc1} the time from the initialization of the call to the moment the related MU leaves the cell. The distributions of variables t_{mci} can be easily demonstrated as [DEL 99]:

$$\Pr ob\{t_{mci} < t\} = \int_{-R}^{R} \Pr ob\{t_{mci} < t|z\}f_i(z)dz \qquad [3.11]$$

With:

$$\text{Prob}\{t_{mci} < t | z\} = \begin{cases} \dfrac{tV_{sat}}{d(z)}\left\{u(t) - u\left[t - \dfrac{d(z)}{V_{sat}}\right]\right\} + u\left[t - \dfrac{d(z)}{V_{sat}}\right], & \text{for } i = 1 \\ u\left[t - \dfrac{d(z)}{V_{sat}}\right], & \text{for } i = 2 \end{cases} \quad [3.12]$$

The *pdf* of the variables t_{mci} is obtained by deriving [3.11] in relation to t. After several algebraic manipulations, we obtain:

$$pdf_{t_{mc1}}(t) = \frac{2}{3}\left\{2 - \frac{t}{\alpha T_m}\right\}\frac{\{u(t) - u(t - \alpha T_m)\}}{\alpha T_m} \quad [3.13]$$

And

$$pdf_{t_{mc2}}(t) = \frac{\{u(t) - u(t - \alpha T_m)\}}{2\alpha T_m} + \frac{1}{2}\delta(t - \alpha T_m) \quad [3.14]$$

With $\delta(t)$ the Dirac delta function.

The distribution function of the random variables t_{mc1} and t_{mc2} is not independent as t_{mc1} can be considered as the residual time in relation to t_{mc2} starting from the arrival instant in the interval [0, t_{mc2}] (*excess life theorem* [CHA 01, EFT 98, WAN 93]). So, from [3.13] and [3.14], we obtain the relation:

$$pdf_{tmc1}(t) = \frac{1 - \int_{\tau=0}^{t} pdf_{t_{mc2}}(\tau)d\tau}{E[t_{mc2}]} \quad [3.15]$$

Given that the *unencumbered call duration* t_d is assumed to be exponentially distributed random variable, the same distribution is valid for the residual duration of the call after a handover request (a memoryless property). As a result, we also designate the residual duration of the call by t_d.

A handover procedure is initialized for an MU with an ongoing call in the cell x as soon as $t_d > t_{mci}$. The probabilities of these events, P_{Hi}, $i=1, 2$, can be obtained as follows:

$$P_{Hi} = \text{Prob}\{t_d > t_{mci}\}$$
$$= \int_0^{+\infty} \text{Prob}\{t_d > t | t_{mci} = t\} \text{pdf}_{t_{mci}}(t) dt$$
$$= \int_0^{+\infty} e^{-(t/T_m)} \text{pdf}_{t_{mci}}(t) dt \qquad [3.16]$$
$$= L\left(\text{pdf}_{t_{mci}}(t)\right)\Big|_{s=1/T_m} \qquad i = 1,2$$

$L(g(t))$ denotes the Laplace transformation of the function $g(t)$ [ROS 83].

By substituting [3.13] and [3.14] in [3.16], we obtain the handover probabilities P_{H1} and P_{H2}:

$$P_{H1}(\alpha) = \frac{2}{3}\left\{P_{h1}(\alpha) + \frac{1 - P_{h1}(\alpha)}{\alpha}\right\}$$
$$\qquad\qquad\qquad\qquad\qquad\qquad [3.17]$$
$$P_{H2}(\alpha) = \frac{P_{h1}(\alpha) + P_{h2}(\alpha)}{2}$$

With:

$$P_{h1}(y) = \frac{1 - e^{-y}}{y}, \qquad P_{h2}(y) = e^{-y} \qquad [3.18]$$

It is worth noting that $P_{hi}(y)$, in $i = 1,2$, represents the probability that a call initializes a handover request for a call (of average duration T_m) in a cell where the relative MU travels a distance (from the moment the call arrives) that is:

– uniformly distributed between *zero* and q for $i = 1$;

– fixed and equal to q for $i=2$.

With $y = q/[V_{sat} T_m]$ [AKR 06].

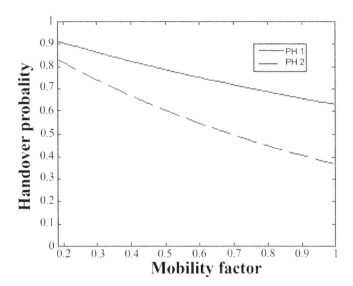

Figure 3.4. Handover probabilities P_{H1} and P_{H2} according to the mobility factor α

It is worth noting that P_{H1} and P_{H2} depend exclusively on the mobility parameter α. Figure 3.4 shows the variations of these two probabilities according to α. It is clear that when α approaches 0 (∞), P_{H1} and P_{H2} get closer to 1 (0).

The *channel holding time* in a cell can be given by [TEK 91]:

$$T_{Hi} = min\ [td\ ,\ t_{mci}] \qquad [3.19]$$

The average value of t_{Hi}, $E[t_{Hi}]$ is given by [TEK 91]:

$$E[t_{Hi}] = T_m(1 - P_{Hi}), \qquad j = 1,2 \qquad [3.20]$$

This equation demonstrates that due to mobility, the average value of channel holding time falls in relation to T_m.

In accordance with the uniform traffic hypothesis, we denote by λ the average arrival rate of new call attempts in a generic cell in the system. The arrival process of handover requests which we are examining here concerns calls that need to be transferred from one cell to another i.e. interspotbeam handover. The arrival process of handover requests in a cell depends on the movement of MUs in relation to cells, on the shape and size of the cells and on the distribution of MUs on the ground. Since all cells are of the same shape and same size (created by the *beamforming* technique), all MUs have the same mobility conditions in relation to the cells (due to the high-speed value of the satellites). Moreover, the distribution of MUs on the ground is uniform and the average arrival rate of handover requests toward a cell is the same for all cells. We obtain a relation between λ_h and λ by applying the condition of flow conservation between the handover requests leaving and entering a given zone [MAR 91]. In particular, we consider the flow conservation equation in horizontal beams with an elementary height dz and length $\sqrt{3}R$ arranged across the cellular network at location z, with z that varies from $z = -R$ to $z = R$ (see Figure 3.3). The elementary arrival rates in a generic beam are $\delta_{na}(z)dz$ and $\delta_h(z)dz$ for new call attempts and handover requests, respectively. The sum of the elementary rates $\delta_h(z)dz$ for beams from $z = -R$ to $z = R$ gives the average arrival rate of handover requests toward a cell:

$$\int_{-R}^{R} \delta_h(z)\,dz = \lambda_h \qquad [3.21]$$

The sum of the regions of the elementary beams from $z = -R$ to $z = R$ is equal to 4/3 of the zone of a cell. Due to the uniform spatial generation of new call arrivals, δ_{na} is independent of z. We, therefore, have:

$$\int_{-R}^{R} \delta_{na}\,dz = \frac{4}{3}\lambda \quad \Rightarrow \quad \delta_{na} = \frac{2}{3R}\lambda. \qquad [3.22]$$

In the analysis that follows, we are going to differentiate between the case $|z| \leq R/2$ and the case $R \geq |z| > R/2$.

1) Case $|z| \leq R/2$: the balanced flow condition is applied to $\delta_{na}\, dz$ and $\delta_h(z)\, dz$ which are linked to a generic elementary beam at the offset z [MAR 91]:

$$\delta_{na} dz (1-P_{b1}) P_{h1}(\alpha) + \delta_h(z) dz (1-P_{b2}) P_{h2}(\alpha) = \delta_h(z) dz \qquad [3.23]$$

Let λ_{hc} denote the average handover requests in the central part of the cell defined by:

$$\lambda_{hc} = \int_{-R/2}^{R/2} \delta_h(z)\, dz. \qquad [3.24]$$

By integrating [3.23] from $z = -R/2$ to $z = R/2$ and by using [3.22] and [3.24], we get:

$$\frac{\lambda_{hc}}{\lambda} = \frac{2}{3} \frac{(1-P_{b1}) P_{h1}(\alpha)}{1-(1-P_{b2}) P_{h2}(\alpha)} \qquad [3.25]$$

2) Case $R \geq |Z| > R/2$: the edges of the cell divide the elementary beam at the offset z into two segments of widths $d(z)$ and $\sqrt{3}R - d(z)$, respectively. We denote by:

$$\alpha_1(z) = \frac{d(z)}{V_{sat} T_m}, \qquad \alpha_2(z) = \frac{\sqrt{3}R - d(z)}{V_{sat} T_m} \qquad [3.26]$$

As a result, $\alpha_1(z) + \alpha_2(z) = \alpha$.

Due to the uniform spatial generation of new calls, the arrival of a new call (in a beam) takes place in the first segment with a probability $d(z)/\sqrt{3}R = \alpha_1/\alpha$ and in the second segment with a probability $(\sqrt{3}R - d(z))/\sqrt{3}R = \alpha_2/\alpha$.

As a result, the arrival of a new call initialized in this beam generates a handover request with a probability

$(1-P_{b1})\left[(\alpha_1/\alpha)P_{h1}(\alpha_1)P_{h2}(\alpha_2)(1-P_{b2})+(\alpha_2/\alpha)P_{h1}(\alpha_2)\right]$, while a call transferred to the beam requires a new handover request with a probability equal to $(1-P_{b2})P_{h2}(\alpha_1)(1-P_{b2})P_{h2}(\alpha_2) = (1-P_{b2})^2 P_{h2}(\alpha)$.

As a result, we arrive at the following flow conservation equation:

$$\delta_{na}dz(1-P_{b1})\left[\frac{\alpha_1}{\alpha}P_{h1}(\alpha_1)P_{h2}(\alpha_2)(1-P_{b2})+\frac{\alpha_2}{\alpha}P_{h1}(\alpha_2)\right]+$$
$$\delta_h(z)dz(1-P_{b2})^2 P_{h2}(\alpha) = \delta_h(z)dz. \quad [3.27]$$

We integrate the two sides of equation [3.27] on the junction – *seam* – of the cellular network, i.e. from $z = -R$ to $z = -R/2$ and from $z = R/2$ to $z = R$. Due to the symmetry of the problem, there is no difference in integrating the two sides of the equation from $z = R/2$ to $z = R$ multiplied by two. Moreover, we divide the result by λ and we obtain $\dfrac{\lambda_h}{\lambda}$:

$$\frac{\lambda_h}{\lambda} = \frac{\lambda_{hc}}{\lambda} + \frac{2}{3}(1-P_{b1})\frac{1-P_{h1}(\alpha)+(1-P_{b2})(P_{h1}(\alpha)-P_{h2}(\alpha))}{\alpha-\alpha(1-P_{b2})^2 P_{h2}(\alpha)} \quad [3.28]$$

Given that:

$$2\int_{R/2}^{R}\delta_h(z)dz = \lambda_h - \lambda_{hc} \quad [3.29]$$

Finally, by substituting [3.25] in [3.28], we get:

$$\frac{\lambda_h}{\lambda} = \frac{2}{3}(1-P_{b1})\left\{\frac{\dfrac{P_{h1}(\alpha)}{1-(1-P_{b2})P_{h2}(\alpha)}+}{\dfrac{1-P_{h1}(\alpha)+(1-P_{b2})(P_{h1}(\alpha)-P_{h2}(\alpha))}{\alpha-\alpha(1-P_{b2})^2 P_{h2}(\alpha)}}\right\} \quad [3.30]$$

According to [MAR 91, EFT 98], the average number of handover requests per call attempt n_h is given by:

$$n_h = \frac{\lambda_h}{\lambda} \frac{handover\ request}{call\ attempt} \qquad [3.31]$$

The parameter n_h is a useful measure for the degree of mobility of the environment. If $P_{b1} = P_{b2} = 0$, the value of n_h given by [3.30] is maximal and equal to $4/(3\ \alpha)$; particularly, in the case of IRIDIUM mobility ($R = 212.5$ km, $V_{sat} = 26\ 600$ km/h), 4.9 handovers are on average required per call with $T_m = 3$ min.

Let us consider a cellular system that is equivalent to the hexagonal cell system in Figure 3.3, but that eliminates the *seams* by using rectangular cells. These cells have the following sizes: $\sqrt{3}R \times 1.5R$ and have the same zone of hexagonal cells. Each MU travels a distance of $\sqrt{3}R$ from one end of the cell to the other. In this case, we use the study presented in [MAR 91] and [KOS 99]: if $P_{b1} = P_{b2} = 0$, $n_h = 1/\alpha$ (3.6 handovers/calls for the IRIDIUM case and $T_m = 3$ min). Therefore, this simplified model underestimates the handover request rate and, as a result, underestimates the probability of blocked calls.

Out of the calls generated, some are blocked (with the probability P_{b1}) and some are admitted into the network (with the probability $(1-P_{b1})$). The average number of handover requests per call admitted in the network n_h' can be linked to the average number of handover requests per attempted call n_h by bearing in mind that the size $1-P_{b1}$ is the average number of calls accepted by the calls generated. As a result, if we multiply n_h' by $1-P_{b1}$, we obtain n_h. The following relation is, therefore, valid:

$$n_h' = \frac{n_h}{1-P_{b1}} \frac{handover\ request}{accepted\ call} \qquad [3.32]$$

It is easy to verify that if an accepted call originates, on average, from n_h' handover request and if each call request can be lost with a probability P_{b2}, the probability of losing an ongoing call is [MAR 91]:

$$P_{drop} = n'_h P_{b2} \qquad [3.33]$$

Finally, the probability of a call failing P_{ns} is given by Maral *et al.* [MAR 91]:

$$P_{ns} = P_{b1} + (1 - P_{b1}) P_{drop} \qquad [3.34]$$

4

Evaluation Parameters Method

4.1. Introduction

Given that the speeds of users and the rotation of the earth are slow in comparison to the speeds of low earth orbit (LEO) satellites, the relative mobile user (MU)-satellite movement is determined by the movement of the satellite, enabling the route of each MU with an ongoing call in the network to be predicted. This, however, requires a global positioning system (GPS) to be integrated into the network so that the precise position of an MU can be located as soon as the call starts. This makes tracking MUs considerably easier, a procedure that is essential for a number of handover request management techniques. Nevertheless, this integration makes designing mobile satellite systems (MSSs) even more complex.

In this research, we will take advantage of the mobility model and its benefits to introduce a new strategy that enables different parameters concerning MUs with ongoing calls to be determined, when the system is not integrated into a global positioning system.

First, we will present the advantages of LEO cellular satellite networks, particularly with regard to analyzing the mobility model. On the basis of this research, we will then introduce a new method for evaluating different parameters concerning MUs with ongoing calls. From this method, we will subsequently derive two new handover request management strategies: a queuing strategy based on *last useful instant* (LUI) that we call Pseudo-LUI (PLUI) and a strategy based on *dynamic channel reservation* (DCR) that we call DCR-*like*.

4.2. The advantages of the LEO MSS mobility model

As explained in the first chapter, the mobility model of LEO satellite systems contains some very important characteristics that make studying LEO systems much simpler than other systems, such as medium earth orbit (MEO) or geostationary earth orbit (GEO).

These characteristics, namely the constant speed and direction of the relative MU-satellite movement, enabled a number of previous studies to determine the different important parameters and information, thus enabling an improvement in network performance.

Indeed, once the position of the MU starting a call has been determined, they can be tracked across the network and their precise location throughout the duration of the call can be established.

This enables us to determine different pieces of information: the position of the MU in the cell, the moment the next handover requests initializes and the maximum waiting time for a call in the overlap zone. First, we will describe the overlap zone related to the mobility model that we have chosen to use in this research.

– *Overlap zone model:*

The maximum queuing time t_{qmax} represents the time required for an MU with an ongoing call to cross the overlap zone.

In general, t_{qmax} is a random variable that depends on the degree of overlap between neighboring spotbeams (characteristics of the antennae and the orbital configuration of the satellites), on the power of the signal, on the propagation conditions (mobile environment: rural, urban, etc.) and on the MU's direction of movement in relation to the cellular plane.

Let us suppose that the satellite ground-tracks are arranged on the earth in a regular hexagonal pattern (with sides R) and have a circular coverage of radius R'. In the literature, the possible values of the relation R'/R vary from 1 to 1.5 [DEL 95a]. Obviously, if the ratio is greater, the overlap zone is wider and the performance of the queuing technique is, therefore, better. We will consider the minimal possible extension of the overlap zone $R = R'$.

Once the position of the MU at the start of the call is defined, an offset z is assigned to this MU in the source cell. Taking into account the regular cellular plane and the suppositions of mobility, the distance $o(z)$ covered by the MU in the overlap zone remains the same for all handover requests, whether they are generated in the source cell or in a transit cell. For an MU crossing the overlap zone at the offset z, the parameter tqmax is given by:

$$t_{qmax} = o(z) / V_{sat} \qquad [4.1]$$

It is worth noting that, according to the suppositions made, the random nature of t_{qmax} depends exclusively on the offset z of the MU traveling through the cellular network.

We will consider here that calls generated in the overlap zone are automatically sent to the destination cell so as to avoid these calls generating handover requests before being served. The term "cell" denotes a region where new calls are managed by the same spotbeam. Researching the overlap zone, therefore, involves a cell with a curvilinear shape, instead of the hexagonal shape that was used previously in the mobility model. The width of the circular cell with radius R and a height z is equal to:

$$r(z) = 2\sqrt{R^2 - z^2} \qquad [4.2]$$

The maximum distance traveled by an MU with an ongoing call in a curvilinear cell before initializing a handover request $h(z)$ is:

$$h(z) = r(z) - o(z) \qquad [4.3]$$

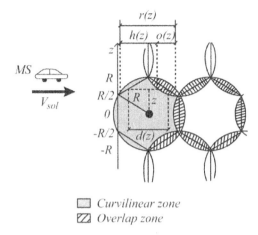

Repetition of Figure 3.2. *Geometry of cellular networks with hexagonal patterns*

With *o(z)* given by:

$$o(z) = \begin{cases} 2\sqrt{R^2 - z^2} - \sqrt{3}R, \\ \sqrt{R^2 - z^2} - \frac{\sqrt{3}}{2}R + \sqrt{R^2 - \left(|z| - \frac{3}{2}R\right)^2}, \end{cases}$$

$$\begin{array}{l} for\ |z| \leq \dfrac{R}{2} \\ for\ R \geq \dfrac{R}{2} \end{array}$$

[4.4]

It is worth noting that the surface of the curvilinear cell is equivalent to that of the hexagonal cell with sides R (i.e. $3\sqrt{3}R^2/2$). Therefore, according to uniform traffic theory, the average arrival rate of new call attempts λ is the same for a hexagonal and a curvilinear cell. On the basis of the new cell shape, we will recalculate the *pdf* of the offset z of a new call attempt in its source cell $f^*(z)$ by using the same approach given in Chapter 3 to derive the *pdf f(z)* of hexagonal cells:

$$f^*(z) = \frac{h(z)}{3\frac{\sqrt{3}}{2}R^2} \quad [4.5]$$

The *pdf* of the offset of an active MU in a transit cell remains uniform, as it was for hexagonal cells. Figure 4.1 compares the distributions $f(z)$ and $f^*(z)$: it can be seen that there is not a great difference between $f^*(z)$ and $f(z)$. Therefore, the analytical results shown in the previous sections can be extended, with a good approximation, to curvilinear cells.

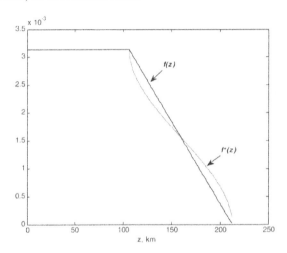

Figure 4.1. *Comparison between the probability density functions $f(z)$ and $f^*(z)$ in the IRIDIUM case*

We will use $f^*(z)$ to determine the average value for the maximum waiting time:

$$E\left[t_{q\max}\right] = \frac{E\left[o(z)\right]}{V_{sat}} \quad [4.6]$$

$$= \frac{1}{V_{sat}} \int_{z=-R}^{z=R} o(z) f^*(z) dz \quad [4.7]$$

$$= \alpha T_m \beta \qquad [4.8]$$

with β equal to:

$$\beta = \frac{4}{9}\left(\frac{\sqrt{3}}{3}\pi - \frac{3}{2}\right) \approx 0.1394. \qquad [4.9]$$

The parameter β depends exclusively on the geometric suppositions of the mobility model and on the overlap zone. For the mobility model in the IRIDIUM case, the average waiting time is close to 7 s.

Let us return now to the parameters that can be determined in the case of a LEO cellular system for an MU with an ongoing call when their initial position is unknown.

Position of the MU in the cell

When an MU starting a call is localized, their position in the cell can be determined with two coordinates z and x. We will designate by z_i and x_i the coordinates of the MU at their initial position, that is their location at the start of the call.

Knowing that the direction and the speed of the relative MU-satellite movement are fixed in relation to the cellular pattern sent to earth by the satellite, the position of the MU in the cell varies exclusively according to the coordinate x and can be determined by:

$$x = x_i + (t_{cl} V_{sat}) \qquad [4.10]$$

t_{cl} being the call duration in the cell.

Maximum queuing time

The maximum time that a call can wait in a queue for a channel to become available in its transit cell represents the period required for the MU to cross the overlap region between their source cell and transit cell; in this region, the call can be served by both cells. This parameter varies exclusively according to the coordinate z and is equal to:

$$t_{qmax} = o(z) / V_{sat} \qquad [4.11]$$

Figure 4.2 shows the variation of t_{qmax} according to z:

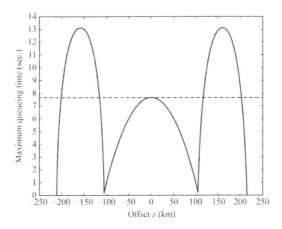

Figure 4.2. *Maximum queuing time according to z for the mobility model*

The moment the next handover request initializes

If the position of an MU with an ongoing call is known, it is possible to estimate the residual time before the latter executes a handover request based on the predictable movement of the MU across the cellular pattern and on the cellular geometry of the network.

We will designate by T_{nxt} the residual time before a call, with coordinates (z, x) in a given cell, starts its next handover request T_{nxt} is given by:

$$T_{nxt} = \frac{h(z) - \left(x + \frac{O(z)}{2}\right)}{V_{sat}} \text{ for } |z| \leq \frac{R}{2}$$

$$T_{nxt} = \frac{h(z) - \left(x - \sqrt{3}R - r(z)\right)}{V_{sat}} \text{ for } R \geq |z| \leq \frac{R}{2} \qquad [4.12]$$

To calculate these parameters, previous research presented in the literature [BOU 03, SOR 99, COO 81, SYS 86] assumed that a positioning system is integrated into the network so that a call can be localized the moment it is initialized. If the initial position of the MU is known it is easy to determine their different parameters.

The location of the MU can be estimated using a GPS that measures the propagation delay and the variation of the Doppler frequency of MU transmissions [BOU 03, HEY 82, ALI 97, ALI 99, ALI 02]. Using time delay measures, we obtain a fixed circle of propagation delay on the earth. Given that the variation of the Doppler frequency relates to the angle between the satellite speed vector and the MU-satellite direction vector, the Doppler measures define a cone by making a fixed angle with the satellite speed vector. The intersection on the earth between the constant circle of propagation delay and the cone identifies two points that represent the possible locations of the MU. A potential solution for resolving this spatial ambiguity is to take another Doppler frequency measure from another visible satellite; this solution imposes a number of additional restrictions on the satellite constellation. Consequently, implementing a positioning system makes implementing the satellite more complicated.

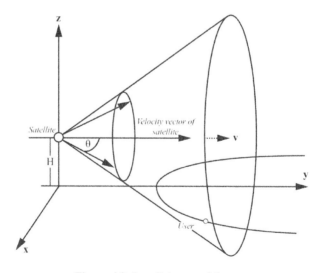

Figure 4.3. *Localizing a mobile user*

In the following section, we will introduce a new method for evaluating parameters that we will call the *evaluation parameters method (EPM)*. EPM [COO 98] demonstrates that it is possible to determine important parameters and information concerning an MU with an ongoing call, even if the location of the latter is unknown, by taking advantage of the predictable relative satellite MU movement and the regular cell pattern of MSSs.

4.3. Evaluation parameters method

The behavior of the relative MU-satellite movement can be predicted as it is determined by the movement of the satellite. Therefore, in LEO MSS systems, the destination cell of a handover request will always be the adjacent cell in the direction of the relative MU-satellite travel.

The time period required for an MU to travel the maximum distance in a cell having initialized a handover request (i.e. the maximum stay time t_{Msj}) is equal to t_{sH} the period separating two consecutive handover request initializations generated by the MU. Using a *timer*, the time period t_{sH} can be estimated by each MU with an ongoing call, while the latter initializes their second handover request.

The maximum stay time t_{Msj} is equal to:

$$t_{Msj} = h(z) / V_{sat}. \qquad [4.13]$$

Different information and parameters can be determined using the value t_{sH} even if the location of the MU is unknown, as will be presented in the following sections.

4.3.1. *Position of the MU in the cell*

Figure 4.4 represents the variation of t_{Msj} ($t_{Msj} = t_{sH}$) according to z for the mobility model that we are considering. It can be seen that the central zone of the cell t_{Msj} has constant values equal to $T_{sM} \approx 50$ s, whereas in the junction zone the values of t_{Msj} are lower than T_{sM} and vary according to the absolute values of z (z and $-z$ give values equal to t_{Msj}).

Consequently, the estimated maximum stay time enables us to determine whether the call belongs to the junction zone or the central zone: if the t_{Msj} is equal to T_{sM} the call is in the central zone, otherwise it belongs to the *junction* region of the cell.

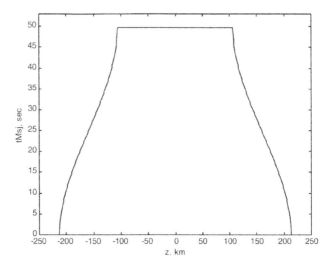

Figure 4.4. *Maximum stay time according to z*

4.3.2. The moment the next handover request initializes

In accordance with the position of the cells and suppositions of mobility, a call belonging to the *central* zone (*junction*) of a cell is transferred toward the *central* zone (*junction*) of the transit cell.

Therefore, for a call belonging to the central zone of the cell, the value of t_{Msj} remains the same in all transit cells that the call will reach as the values of t_{Msj} are constant in this zone.

For calls belonging to the *junction* zone, this time period changes from cell to cell. However, it is easy to determine the next value of t_{Msj} by using the previous value in the previous cell:

Let us suppose that an MU with an ongoing call at the offset z in the *junction* zone of a cell and with a maximum stay time t_{Msja} is moved to another cell. The new offset z_n (Figure 4.5) related to this call in the coordinate system of the transit cell is equal to:

$$|z_n| = \frac{3}{2}R - |z| \qquad [4.14]$$

Therefore, the value of the maximum stay time in the new cell t_{Msjb} is given by:

$$t_{Msjb} = h(z_n) / V_{sat} \qquad [4.15]$$

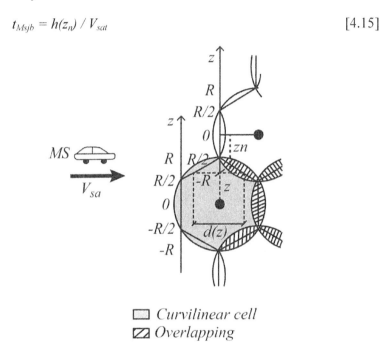

☐ *Curvilinear cell*
▨ *Overlapping*

Figure 4.5. *New offset z_n in the transit cell*

In representing the variation of t_{Msja}, t_{Msjb} and $t_{Msja} + t_{Msjb}$ according to $|z|$ in the junction zone of a cell (the result is given in Figure 4.6), it is noted that the sum of t_{Msja} and t_{Msjb} is constant and equal to T_{sM}. Therefore, for every call in the junction zone initializing its second handover request, the maximum stay time in the neighboring cell is given by:

$$t_{Msjb} = T_{sM} - t_{Msja} \qquad [4.16]$$

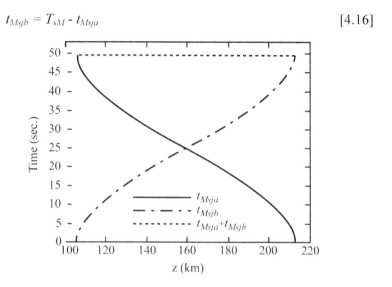

Figure 4.6. *Maximum stay time in the source and transit cells according to |z|*

This time period t_{Msj} enables the system to precisely estimate the residual time before the call initializes a new handover request: a timer is started while a call initializes a handover request. The time that passes from this moment, which is designated by t_{hr}, is therefore used to evaluate the time that remains before the call implements its next handover request:

$$T_{nxt} = t_{Msj} - t_{hr}.$$

4.3.3. *Maximum queuing time*

Figure 4.7 represents the variation of t_{Msj} and t_{qmax} according to z for the mobility model that we are considering. It is noted that for the *junction* zone each value of t_{Msj} has a value equivalent to t_{qmax}. It is also noted that the values equal to t_{Msj} in the different zones (for z and $-z$) have the same value of t_{qmax}. This is due to the symmetry of the system topology.

Let us represent the variation of t_{qmax} according to t_{Msj} using relations [4.11] and [4.13] [DEL 94]; this result is presented in Figure 4.8. In reference to this result, it is possible for the calls belonging to the *junction* zone to determine the maximum queuing time from the moment the second handover request is initialized using the value of t_{Msj}. For calls initialized in the central zone of a cell, the information available is that $t_{q\max} \leq 7.706$ sec (see Figure 4.2).

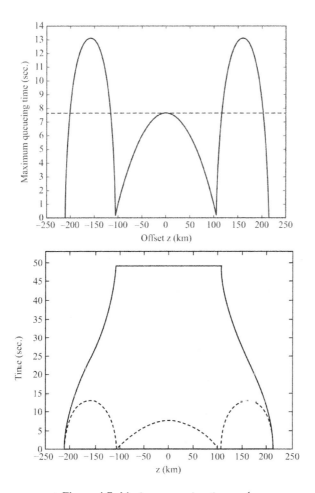

Figure 4.7. *Maximum queuing time and stay time according to z*

Due to the regular cell arrangement and the mobility hypotheses, it is noted that the maximum queuing time for MUs remains the same for every handover request, whether they are from the source cell or a transit cell [BOU 03].

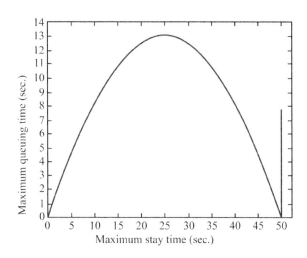

Figure 4.8. *Maximum queuing time according to the maximum stay time*

Figure 4.7 represents the variation of t_{sH} and t_{qmax} according to z; it is noted that with the exception of the central region of the cells, each value of t_{sH} has a corresponding value of t_{qmax}. Using these values, Figure 4.8 gives the variation of t_{qmax} according to t_{sH}. It must be noted that due to the symmetry of the system topology, the same values of t_{sH} in different regions (z and $-z$) correspond to the same values of t_{qmax}. The result is that it is possible to determine the values of t_{qmax} by using the values of t_{sH}.

Indeed, Figure 4.8 represents the graph of the function $-ax^2+bx$ (*see* Figure 4.9). Using the values of t_{qmax} and t_{sH}, it is possible to obtain the values of a and b:

$$t_{qmax}(t_{sH}) = -a(t_{sH})^2 + b(t_{sH}). \qquad a \approx 0.0212 \quad b \approx 1.0551 \qquad [4.17]$$

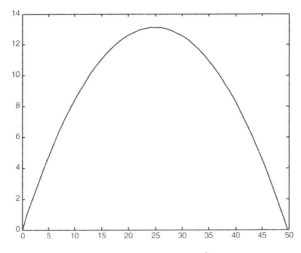

Figure 4.9. *Function* $y = -a x^2 + b x$

On the basis of this method, we will present two guaranteed priority strategies for handover requests, PLUI and DCR-*like* in the following sections.

4.4. Pseudo-last useful instant queuing strategy

4.4.1. *Putting handover requests in a queue*

Let us consider an MU with an ongoing call leaving a cell x and moving toward an adjacent cell y; there is a zone where this MU can receive signal with an acceptable level of power from the two cells x and y; this zone is commonly called the overlap zone. The time period t_{qmax} during which an MU crosses the overlap zone can be used to put their handover request in a queue if no channel is free in cell y. To meet International Telecommunication Union (ITU) quality of service requirements [ITU 95], it is essential to adopt a management strategy for calls with handover requests.

4.4.2. *Handover request management*

Let us suppose that a handover procedure initializes as soon as an active MU crosses into the overlap zone between the two cells x and y.

We will denote by $\Lambda(y)$ the group of channels available in y. The handover request is served according to the following predispositions:

1) if $\Lambda(y) \neq \phi$, the call is immediately switched, a new channel is allocated to the MU in y and the old channel in x is freed;

2) if $\Lambda(y) = \phi$, l the handover request is put in a queue to wait for a channel to become available in the transit cell y. In the meantime, the call is served by the cell x. A handover request leaves the queue in the event of one of the following:

i) *The handover procedure is successful:* the call is switched before it ends or its waiting time expires.

ii) *The handover request is rejected:* the call ends before it is switched or its waiting time expires.

iii) *The handover procedure fails and the call is lost:* the handover request was not served during t_{qmax} and the call did not end before the waiting time expired.

4.4.3. LUI queuing strategy

Since the 1980s, a number of queuing strategies for handover requests have been suggested, first for land cellular systems and later for satellite systems.

Introduced in 1999 by Del Re *et al.*, the LUI queuing strategy is considered an ideal strategy for mobile satellites [DEL 99] as it is based on a precise evaluation of handover requests' queuing times; the system makes a precise evaluation of the residual waiting times t_{qmax} of all the requests entering in a queue.

A new request is attributed a position in the queue before (after) all the handover requests with a higher (lower) residual waiting time than their waiting time t_{qmax} [KIA 11], so that the most urgent handover requests are served first as soon as a channel is freed in the transit cell.

The LUI technique can be practically implemented in LEO-MSSs by using an adapted positioning system that evaluates the position of the MU at the start of the call and follows them throughout the communication. It is worth noting that once the position of the MU is known, they can be tracked by estimating the variations in their position based on the satellite time standard. The time period t_{qmax} can, therefore, be obtained from equation [4.11].

The effectiveness of this discipline in relation to first in first out (FIFO) – the most popular queuing strategy – essentially depends on the degree distribution of tqmax around its average value. If this value is close to a determined value then the LUI discipline is close to FIFO. Note that the distribution of tqmax depends on the suppositions taken from the mobility model and the overlap zone. The effectiveness of this discipline also depends on the mobility coefficient.

Nevertheless, implementing a positioning system makes implementing a satellite system more complex. To reach a compromise between the simplicity of the FIFO strategy and the effectiveness of the LUI strategy, in the following section we will present an alternative PLUI strategy [KIA 08] based on the EPM presented in the previous section.

4.4.4. *Pseudo-LUI queuing strategy*

The aim of this strategy is to evaluate an approximate maximum waiting time for an active user entering in a queue through the use of the EPM rather than integrating a positioning system, the major disadvantage of the LUI strategy.

Indeed, the regular arrangement of the cells and the mobility characteristics mean that the distance $o(z)$ that an MU travels in the overlap zone remains the same for any handover request, whether it is generated from the source cell or a transit cell. Therefore, it is sufficient to estimate the maximum queuing time for an ongoing call once and then use it again in any call from the second initialization of a handover request.

By applying the EPM, it is possible to estimate t_{sH} (maximum stay time) for each MU with an ongoing call.

Therefore, in the PLUI strategy, the system evaluates the maximum stay time t_{sH} for each call initializing its second handover request and uses it to estimate the maximum queuing time that the call could spend waiting for a free channel. This value is recorded so it can be used if the call requires another handover procedure.

For the problem of the central region (for which t_{sH} remains the same for different values of t_{qmax}), we suggest considering the maximum time t_{mxqp} that these calls spent waiting on previous occasions and using this time to estimate the real maximum queuing time with the following formula:

$$t_{qmax} \cong t_{mxqp} + \left(\left(t_{mxc} - t_{mxqp} \right) / 5 \right) \qquad [4.18]$$

t_{mxc} is the maximum value of t_{qmax} in this region (see Figure 4.2). This period is used to order the following handover requests in a queue according to the LUI strategy.

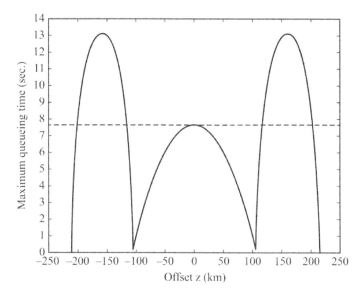

It is important to note that these calls represent a minority of queuing requests. Indeed, in [MAR 97], Markoulidakis *et al.* provide an estimation of the average number of handover requests per call for the same mobility model. Their research shows that the average number of handover requests initialized by a call from the central region of a cell is close to half of that of requests initialized by a call in the junction zones. It is, therefore, supposed that this problem will only have a limited impact on the effectiveness of the strategy proposed.

New calls initializing their first handover requests are ordered in a queue according to the FIFO strategy and before all the other requests. Priority in the queue is, therefore, given to calls initializing their first handover requests.

4.5. Guard channel strategy: dynamic channel reservation-*like*

4.5.1. *Dynamic channel reservation technique [DEL 95, KIA 11]*

Handover with guard channel reservation strategies gives priority of success to handover requests by reserving a certain number of channels exclusively for them (adjusting the number of channels can be fixed or dynamic). The other channels are used for new calls as well as for handover requests. This reduces the probability of losing calls during a handover procedure, whereas, however, the probability of new call attempts being rejected increases as fewer channels are available to them. Selecting the number of guard channels, therefore, requires careful consideration.

In this research, we will adopt the IRIDIUM mobility model (used in [GRU 91]), where the cells are supposed to be rectangular shaped and with a width of $R = 425\ km$ (see Figure 4.10).

DCR strategy reserves channels dynamically according to a parameter system called channel reservation number (CRN). This system records the moment each ongoing call enters the overlap zone and tracks how long it is there. The longer it is there, the lower the probability the call requires a handover to the neighboring cell. The CRN parameter system is actualized each time an event occurs, i.e. call arrivals, handover requests or call end. CRN is given according to P_{hj} and w_j:

$$CRN = \sum_{j=1}^{C} w_j P_{hj} \qquad [4.19]$$

where C is the total number of channels in each cell and P_{hj} is the probability that an ongoing call j will initialize a handover request. It is equal to:

$$P_{hj} = \exp\left(-\frac{t_{Msj}}{T_m}\right) \qquad [4.20]$$

Where t_{Msj} the stay time of the call in the current cell.

This probability must be recalculated when the call implements a handover.

Each probability is *weighted* by the position factor w_j which is the position of the MU x_j divided by the radius of the cell R, i.e. $w_j = x_j/R$. This factor is used to determine the urgency of reserving a corresponding channel.

Channels are reserved in a cell according to the value of the CRN parameter system in the previous cell. CRN is rounded off to the closest number and is then used as the number of channels reserved in the next cell.

This process is illustrated in the graph shown in Figure 4.10. When a new call arrives, the traffic conditions in the cell *1* change, and therefore the CRN of this cell is recalculated and the cell *2* will try its hardest to reserve the number of channels according to the new value of CRN. When a handover is implemented from cell B toward cell C, the traffic conditions of cell B and cell C change, and therefore recalculating the CRNs of cells B and C is indispensable. In the meantime, and due to the fact that there is a free channel in cell B, there is the possibility that this channel can be reserved for any potential handover requests from cell A if cell A does not have enough channels reserved. As a result, cells B, C and D will adjust their reserved channels according to the different values of CRN. When a call ends in cell a, its CRN will be recalculated and cells a and b will adjust their reserved channels.

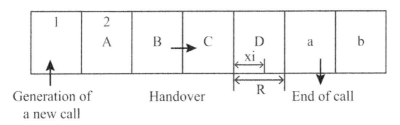

Figure 4.10. *Dynamic channel reservation (DCR)*

In [RES 95], it is assumed that the position of the user is known at the start of the call as a positioning system is integrated into the LEO MSS. This is important and even indispensable for the DCR strategy since the CRN parameter system is calculated according to the position of calls in the cell. This integration is, however, seen as complex. To avoid it, we are putting forward a replacement DCR strategy based on the EPM that we call DCR-*like* [KIA 11].

4.5.2. *Dynamic channel reservation DCR-like technique*

In this section, the DCR strategy will be used in different conditions: the position of the MU is unknown and the moments of consecutive handover requests are determined using the EPM.

Therefore, it is only at the moment the second handover request of an ongoing call initializes that the maximum stay time t_{Msj} is estimated $(t_{Msj} = t_{sH})$ and then used to calculate the probability that the call will initialize another handover request P_{hj}. The distance traveled in the new transit cell is calculated according to t_{hr}, time spent in the cell from the moment it switched: $x_i = V_{sat} t_{hr}$.

At the moment of the first handover request, it is impossible to discern whether the call is in the junction zone or in the central zone. It is, therefore, necessary to wait for the initialization of the second handover request to be able to determine the value of the maximum stay time. To simplify matters, we only consider the central zone of the cells. In this zone, t_{Msj} has a constant value and, as a result, the system determines the position x of a call at the initialization of the first handover request. It is as if the cells were arranged in a rectangular pattern.

When a new call initializes, given that the system is incapable of determining the maximum stay time of the MU, the *channel locking mechanism* strategy [MAR 98] is adopted: for each new call, a channel is allocated in the source cell and a second is *closed* in the next transit cell for its first handover request. The new calls, therefore, have a *guaranteed handover*. Consequently, in this case, the DCR strategy is applied exclusively for calls that are not in their source cell.

Therefore, following the DCR-*like* strategy, when a new call arrives in cell 1 the system tries to allocate a channel in cell 1 and to block another in cell 2. A new call can only be accepted if two channels are available in the two cells.

If a new call terminates in cell 1, this channel as well as the channel blocked in cell 2 for its first handover is freed, resulting in a change in the traffic conditions. The CRNs of cells are, therefore, recalculated as explained for the DCR technique.

5

Analytical Study

5.1. Introduction

Due to the mathematic complexity of analytical studies for dynamic channel allocation systems, the majority of research in this area has been conducted by simulation [BOU 01, WAN 93, DEL 94, DEL 95, DEL 96, DEL 97, BOG 01, DEL 93, CHE 96, LIL 05].

This chapter will present an analytical study of first in first out (FIFO), last useful instant (LUI) and pseudo-LUI (PLUI) queuing strategies for fixed channel allocation (FCA) as well as an analytical study for fixed channel reservation (FCR) and FCR-*like* strategies. This study will be used to evaluate the simulation that will be presented in Chapter 6 by comparing the analytical and simulation results for the fixed cases.

5.2. An analysis of FCA-QH with different queuing strategies

This section will consider an analytical approach to evaluating the performance of FCA queuing handover (QH). The approach is based on the following suppositions and approximations:

– S channels are assigned to a cell according to equation [3.1];

– new call arrivals and handover requests are two independent Poisson processes with the average rates λ and λ_h, respectively. λ_h relates to the next λ [3.30];

– the static distribution of channel *holding time* in a cell for new call arrivals and handover requests is approximated by an exponential distribution with the average value $1/\mu$, given by:

$$\frac{1}{\mu} = \frac{\lambda(1-P_{b1})}{\lambda(1-P_{b1})+\lambda_h(1-P_{b2})} E[t_{H1}] + \frac{\lambda_h(1-P_{b2})}{\lambda(1-P_{b1})+\lambda_h(1-P_{b2})} E[t_{H2}] \quad [5.1]$$

where $E(t_{H1})$ and $E(t_{H2})$ are obtained from [3.20];

– the maximum queuing time is approximated by an exponentially distributed random variable, with an average value $1/\mu\omega$, given by [4.8];

– an infinite queue.

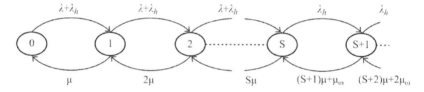

Figure 5.1. *Queuing system for an FCA-QH Markov chain*

In light of the above, each cell can be "modeled" as an *M/M/S* queuing system with non-homogeneous arrival rates [MAR 91, XU 00, TEK 91] (*M*: Poisson arrival process/*M*: exponentially distributed service time/*S*: number of channels assigned per cell), as shown in Figure 5.1. The state of this system is given by the sum of the number of calls in service and the number of handover requests in the queue. When the system is at a state n less than S, the total arrival rate is $\lambda+\lambda_h$; however, if the state is higher than or equal to S (i.e. all the channels are held), the *gross* rate of arrival becomes λ_h. Moreover, we took into account that a call can end up in the overlap zone before being served and that its queuing time expires [COO 98].

Consequently, the Markov chain in Figure 5.1 contains the additional "*death*" rate $i\mu$ for the states $S+i$, with $I = 1, 2,...$ This Markov chain is valid for FIFO, LUI and PLUI strategies.

The probability of the state n, P_n, is:

$$P_n = \begin{cases} \dfrac{(\lambda+\lambda_h)^n}{n!\mu^n} P_0, & 1 \leq n \leq S-1 \\ \dfrac{(\lambda+\lambda_h)^S \lambda_h^{n-S}}{S!\mu^S \prod_{j=1}^{n-S}((S+j)\mu + j\mu_\omega)} P_0, & n \geq S \end{cases} \quad [5.2]$$

with:

$$P_0 = \left\{ \sum_{n=0}^{S-1} \left[\dfrac{(\lambda+\lambda_h)^n}{n!\mu^n} \right] + \sum_{n=S}^{\infty} \left[\dfrac{(\lambda+\lambda_h)^S \lambda_h^{n-S}}{S!\mu^S \prod_{j=1}^{n-S}((S+j)\mu + j\mu_\omega)} \right] \right\}^{-1} \quad [5.3]$$

New calls arriving in the system are blocked when all available channels are held, i.e. when the queuing system is at state $n \geq S$. Therefore, P_{b1} is equal to:

$$P_{b1} = \sum_{n=S}^{\infty} P_n \quad [5.4]$$

The probability P_{b1} does not depend on the queuing discipline used for handover requests. This was verified with simulations (see Chapter 7). P_{b2}, meanwhile, depends on the chosen queuing policy.

With the FIFO strategy, P_{b2} can be conducted by a study presented in [AKY 99] and [HON 86] and taking into consideration that:

– P_{b2} must contain, as a multiplying factor, the probability P_{uh} that the call with a handover request in a queue does not end before t_{qmax} expires. Given that we have exponential distributions of the maximum queuing time and the channel holding time, P_{uh}, is:

$$P_{uh} = \frac{\mu_\omega}{\mu + \mu_\omega} \quad [5.5]$$

– the probabilities of the states are given by [5.2] and [5.3];

– we consider the additional "*departure*" rates $i\mu$ for the states $S+i$ with $I = 1, 2,...$ due to the calls that end in the overlap zone before their handover procedures are completed.

Therefore, we have:

$$P_{b2} = \frac{\mu_\omega}{\mu + \mu_\omega} \sum_{n=S}^{\infty} P_n \left\{ 1 - \frac{S\mu}{S\mu + \mu_\omega} \right.$$
$$\left. \cdot \prod_{j=1}^{n-S} \left[1 - \frac{\mu_\omega}{S\mu + \mu_\omega} \left(\frac{\mu_\omega}{\mu + \mu_\omega} \frac{1}{2} \right)^j \right] \right\} \quad [5.6]$$

For more details, please see Appendix 3.

With the LUI strategy, each handover request reaches the top of the queue unless the request leaves the queue because the call is terminated. Therefore, only the requests at the top of the queue risk failing. As a result, the probability of failure for a handover request that initially enters a queue at the position $n > S$ does not depend on n.

We designate the probability of a handover request failing by $P_{b2|S}$. Therefore, using [5.4], we then obtain [BOU 01]:

$$P_{b2} = P_{b2|s} P_{b1} \quad [5.7]$$

$P_{b2|S}$ takes into account two joint and independent events:

1) Call A whose handover request is at the top of the queue does not end before its maximum queuing time expires. The probability of this event is P_{uh}.

2) None of the S channels are released in the transit cell of A before its maximum queuing time expires. The probability of this event is denoted by P_f. According to the exponential distributions of maximum queuing times and channel holding times, we have:

$$P_f = \frac{\mu_\omega}{S\mu + \mu_\omega} \qquad [5.8]$$

To conclude, in the case of LUI, P_{b2} is given by:

$$P_{b2} = P_{b1} \frac{\mu_\omega}{\mu + \mu_\omega} \frac{\mu_\omega}{S\mu + \mu_\omega} \qquad [5.9]$$

For the PLUI strategy, the problem discussed in the previous section concerning calls belonging to the central region is ignored.

We also suppose that the nth request that enters a queue has already been switched and needs to be switched again.

Since the calls ranked in the queue according to the FIFO strategy are a priority and since only the request at the top of the queue risks failing in calls ranked according to the LUI strategy, the probability that the nth request that enters a queue will be blocked is given by:

$$P_{b2} = \frac{\mu_\omega}{\mu + \mu_\omega} \sum_{n=S}^{\infty} P_n \sum_{nf=0}^{n-S} \left\{ \left(1 - \frac{S\mu}{S\mu + \mu_\omega}\right) \right. \\ \left. \cdot \prod_{j=1}^{nf} \left[1 - \frac{\mu_\omega}{S\mu + \mu_\omega}\left(\frac{\mu_\omega}{\mu + \mu_\omega}\frac{1}{2}\right)^j\right] \right\} \left(\frac{\mu_\omega}{S\mu + \mu_\omega}\right)^{\left(1-0^{n-s-nf}\right)} \qquad [5.10]$$

Note that a recursive approach is required to calculate P_{b1} and P_{b2} (for the three queuing disciplines) according to λ, since λ_h depends on both (P_{b1} and P_{b2}). To speed up convergence, the iterative method is based on the parameter $n_h = \lambda_h/\lambda$ according to P_{b1} and P_{b2} [DEL 99].

We start the iteration with the value of n_h obtained from [3.30] and [3.31] with $P_{b1} = P_{b2} = 0$ (this is the maximum value of n_h, which decreases for the ascending values of P_{b1} and P_{b2}). With this value for n_h, μ and P_n, $n = 0, 1, ...$, are calculated from [5.1]–[5.3]. These values are used to determine P_{b1} and P_{b2} and therefore the new value of n_h. This value is approximated with the value used in the previous step. A new iteration starts with this average value of n_h. The iterative method is stopped when the relative difference between the two values of consecutive n_h is lower than a given threshold (e.g. 10^{-3}). Finally, P_{ns} is derived from [3.34]:

$$P_{ns} = P_{b1} + (1 - P_{b1})\frac{n_h}{1 - P_{b1}} P_{b2} \qquad [5.11]$$

Figure 5.2 gives the analytical results concerning FCA-QH with FIFO, LUI and PLUI strategies in the IRIDIUM case with $S = 10$. It has been noted that there is no significant difference between the performances of the three queuing strategies. This result can be explained by the fact that the MU spends a small amount of time in the overlap zone (7 seconds on average) and each cell manages its own queue so that only a limited number of calls are ranked together. As a consequence, there will not be a great difference between serving the most urgent request (LUI, PLUI) and serving the oldest request (FIFO).

With regards to dynamic channel allocation, LUI and PLUI strategies will have more impact on the performance of the network, as will be shown in the following section. For this case, the results are obtained by simulation, due to the great complexity of the theoretical study [BOU 01].

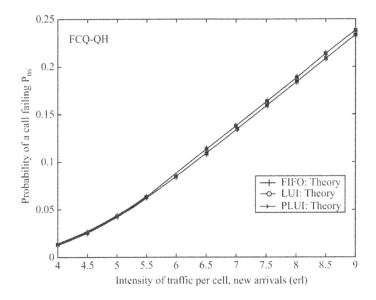

Figure 5.2. *Theoretical comparison between the performance of FIFO, LUI and PLUI queuing strategies for the FCA-QH case according to P_{ns} (IRIDIUM case, S = 10)*

5.3. Analytical study of FCR and FCR-*like*

Given the mathematical difficulty of analyzing dynamic channel reservation (DCR) and DCR-*like*, we will continue here by analyzing FCR and FCR-*like* strategies: a number of fixed channels $C_h = 2$ are reserved for handover requests and the *locking channel mechanism* is used for new calls in the case of FCR-*like*.

5.3.1. *An analysis of FCR*

The probabilities that a call in its source cell, or in its transit cell, initializes a handover request are, respectively:

$$P_{h1} = \frac{1-e^{-\alpha}}{\alpha}, \qquad P_{h2} = e^{-\alpha} \qquad [5.12]$$

where α is the parameter without dimension that characterizes the mobility of the user (for the mobility model considered in Chapter 4 for DCR): $\alpha = \dfrac{R}{T_m V_{sat}}$.

The analytical approach is based on almost the same approximations and suppositions made in the previous section, including:

– new call arrivals and handover requests are two independent Poisson processes with the average rates λ and λ_h per cell with $n_h = \lambda_h / \lambda$;

– an infinite population of users and uniform traffic;

– the channel holding time for new calls and handover requests is approximated by an exponential distribution with an average $1/\mu$ expressed by:

$$\frac{1}{\mu} = \frac{\lambda(1-P_{b1})}{\lambda(1-P_{b1})+\lambda_h(1-P_{b2})} E(t_{h1}) + \frac{\lambda_h(1-P_{b2})}{\lambda(1-P_{b1})+\lambda_h(1-P_{b2})} E(t_{h2}) \quad [5.13]$$

$E[t_{h1}]$ and $E[t_{h2}]$ are obtained from:

$$E[t_{hi}] = T_m(1-P_{hi}), \ i=1,2 \quad [5.14]$$

By applying the flow conservation equation, we have:

$$n_h = \frac{(1-P_{b1})P_{h1}}{1-(1-P_{b2})P_{h2}} \quad [5.15]$$

In FCR, priority is given to handover requests: C_h channels are allocated exclusively to handover requests out of the C channels available in each cell. The remaining channels $C-C_h$ are shared between new calls and handover requests. Therefore, a new call is blocked if the number of channels available in the cell is lower than or equal to C_h at the moment the call initializes. A handover request only fails if no channel is available in the destination cell.

Analytical Study

Each cell can be modeled as an *M/M/C/C_h* queuing system with non-homogeneous arrival rates. The state of the system is defined as being the number of calls in service. The state-transition diagram is shown in the following figure.

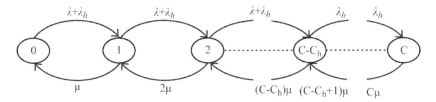

Figure 5.3. *State-transition diagram*

The probabilities of the states are given by:

$$P_j = \begin{cases} \dfrac{\lambda + \lambda_h}{j\mu} P_{j-1}, & 1 \leq j \leq C - C_h \\[6pt] \dfrac{\lambda_h}{j\mu} P_{j-1} & C - C_h + 1 \leq j \leq C \end{cases} \quad s \quad [5.16]$$

By using recursive equations with the normalization condition:

$$\sum_{j=0}^{C} P_j = 1 \qquad [5.17]$$

We obtain:

$$P_0 = \left[\sum_{k=0}^{C-C_h} \dfrac{(\lambda + \lambda_h)^k}{k!\mu^k} + \sum_{k=C-C_h}^{C} \dfrac{(\lambda + \lambda_h)^{C-C_h} \lambda_h^{k-(C-C_h)}}{k!\mu^k} \right]^{-1} \qquad [5.18]$$

$$P_j = \begin{cases} \dfrac{(\lambda + \lambda_h)^j}{j!\mu^j} P_0, & 1 \leq j \leq C - C_h \\[6pt] \dfrac{(\lambda + \lambda_h)^{C-C_h} \lambda_h^{j-(C-C_h)}}{j!\mu^j} P_0, & C - C_h + 1 \leq j \leq C \end{cases} \qquad [5.19]$$

Therefore, the probability that a handover request fails, P_{b2}, is equal to the probability that the system is at state C:

$$P_{b2} = \frac{(\lambda + \lambda_h)^{C-C_h} \lambda_h^{C_h}}{C! \mu^C} P_0 \qquad [5.20]$$

and the probability of finding all the channels in a cell held, P_{b1}, is:

$$P_{b1} = \sum_{j=C-C_h}^{C} P_j \qquad [5.21]$$

The probabilities of losing a call and call failure P_{drp} and P_{ns}, respectively, are given by:

$$P_{drp} = \frac{P_{b2} P_{h1}}{1-(P_{h2}(1-P_{b2}))} \qquad [5.22]$$

$$P_{ns} = P_{b1} + (1-P_{b1}) P_{drp} \qquad [5.23]$$

5.3.2. *An analysis of FCR-like*

We consider that new calls arrive at the system following a Poisson process, the channel holding time is exponentially distributed, traffic is uniform and there is an infinite population of users.

The traffic components in a given cell are:

– λ_{Gn}: arrival rate of new calls;

– λ_h: arrival rate of handover requests, $\lambda_h = \lambda_{Fh} + \lambda_{Sh}$:

- with λ_{Fh} arrival rate of calls initializing their first handover request and,

- λ_{Sh}: arrival rate of other handover requests.

Note that a given cell receives channel allocation requests due to new communication attempts generated by users in the cell itself and users in the previous cell.

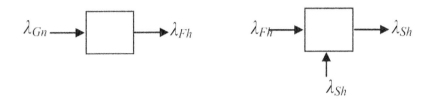

Figure 5.4. *Flow of calls entering and leaving*

To determine the expression of the arrival rate of handover requests according to the arrival rate of new calls and the probability of blocking P_b (i.e. the probability that a request to establish a connection finds all channels held in a cell), handover requests are assumed to arrive in a cell following a Poisson process, which is independent of the arrival process of new calls and subject to the balanced flow condition in a cell between handovers that are entering and leaving [MAR 98]. This condition of balance will be applied separately for new calls and handover requests taking into consideration that:

– a new call attempt is accepted only if a channel is free in the source cell and in the first transit cell. Given that all cells are identical and have the same traffic parameters, the probability of simultaneously finding a free channel in each of the two cells is equal to $(1-P_b)^2$, and therefore the probability of blocking new call attempts is: $P_{b1} = 1 - (1-P_b)^2$;

– the first handover request for each accepted call is never blocked; therefore, the probability that the first handover request succeeds is equal to 1;

– new calls that have been initialized generate the first handover requests and the latter with calls that have already been switched generate other handover requests (see Figure 5.4). Therefore, we have:

$$\lambda_{Gn}(1-P_b)^2 P_{h1} = \lambda_{Fh} \qquad [5.24]$$

$$\lambda_{Sh}(1-P_{b2})P_{h2} + \lambda_{Fh}P_{h2} = \lambda_{Sh} \qquad [5.25]$$

Therefore, we obtain:

$$\lambda_{Sh} = \frac{P_{h2}}{1-(1-P_{b2})P_{h2}} \lambda_{Fh} \qquad [5.26]$$

$$\Rightarrow \lambda_{Sh} = \frac{P_{h1}P_{h2}(1-P_b)^2}{1-(1-P_{b2})P_{h2}} \lambda_{Gn} \qquad [5.27]$$

From the aforementioned relations, the arrival rate of handover requests $\lambda_h = \lambda_{Fh} + \lambda_{Sh}$ is given by:

$$\lambda_h = \frac{P_{h1}(1-P_b)^2(1+P_{h2}P_{b2})}{1-(1-P_{b2})P_{h2}} \lambda_{Gn} \qquad [5.28]$$

The determination of the channel holding time in a cell by an exponential distribution with a rate μ is as in [QUI 04, DEL 99, SYS 86]. We consider the average channel holding time in a cell for different types of calls. Consequently, each average value is weighted with its probability of occurrence. These contributions are therefore added to obtain the average channel holding time. The following definitions are characteristic:

P_1: the probability that a channel is held (or closed) by a new call of the cell being considered (or the previous cell).

P_2: the probability that a channel is held by a switched call (a call that is not in its first transit cell).

$E\ [T_{h1}]$: average value of the channel holding time in the source cell.

$E\ [T_{h2}]$: average value of the channel holding time in a transit cell.

$E\ [T_{h3}]$: average value of the channel locking time in the first transit cell. Note that it includes both the waiting time and the active time.

As such, the average channel waiting time in a cell is given by:

$$\frac{1}{\mu} = P_1 E[T_{h1}] + P_2 E[T_{h2}] + P_1 E[T_{h3}] \qquad [5.29]$$

with P_1 and P_2 equal to:

$$P_1 = \frac{\lambda_{Gn}(1-P_b)^2}{\Lambda} \qquad [5.30]$$

$$P_2 = \frac{\lambda_{SH}(1-P_{b2})}{\Lambda} \qquad [5.31]$$

Λ is the average rate of the total traffic transported:

$$\Lambda = 2\lambda_{Gn}(1-P_b)^2 + \lambda_{Sh}(1-P_{b2}) + \lambda_{Fh} \qquad [5.32]$$

The average holding times are:

$$E[T_{h1}] = T_m[1-P_{h1}] \qquad [5.33]$$

$$E[T_{h2}] = T_m[1-P_{h2}] \qquad [5.34]$$

$$E[T_{h3}] = T_m[1-P_{h1}P_{h2}] \qquad [5.35]$$

And so, we have:

$$\frac{1}{\mu} = \frac{\lambda_{Gn}(1-P_b)^2}{\Lambda}\left(T_m[1-P_{h1}] + T_m[1-P_{h1}P_{h2}]\right) + \frac{\lambda_{SH}(1-P_{b2})}{\Lambda} T_m[1-P_{h2}] \qquad [5.36]$$

The total average rate of call arrivals in a cell is the sum of the different average arrival rates:

$$\lambda_t = 2\lambda_{Gn}(1 - P_b) + \lambda_h \quad [5.37]$$

This expression takes the two contributions with a relation to the new calls into consideration: the contribution that comes from the cell itself and the contribution that comes from the previous cell, each of which is conditioned by the requirement of having a free channel in the previous or next cell.

In light of the above, each cell can be modeled as an M/M/C queuing system with arrival rates of new calls and handover requests λ_{Gn} and λ_h, respectively [MAR 98]. The number of calls in service represents the state of the queuing system. The probability of the state j, P_j, is given by:

$$P_j = \begin{cases} \dfrac{\lambda_t^j}{j!\mu^j} P_0, & 1 \leq j \leq C - C_h \\[1em] \dfrac{\lambda_t^{(C-C_h)} \lambda_h^{j-(C-C_h)}}{j!\mu^j} P_0, & C - C_h + 1 \leq j \leq C \end{cases} \quad [5.38]$$

where the probability P_0 that the system is inactive is:

$$P_0 = \left[\sum_{k=0}^{C-C_h} \frac{\lambda_t^k}{k!\mu^k} + \sum_{k=C-C_h+1}^{C} \frac{\lambda_t^{C-C_h} \lambda_h^{k-(C-C_h)}}{k!\mu^k} \right]^{-1} \quad [5.39]$$

The probability P_b of finding all the channels in a cell held is:

$$P_b = \sum_{j=C-C_h}^{C} P_j$$

A new call is accepted only if a channel is inactive in the source cell and the first transit cell. Consequently, the probability of a new call being blocked is given by:

$$P_{b1} = 1 - (1 - P_b)^2 = 2P_b - P_b^2 \quad [5.40]$$

The probability of a handover request failing is:

$$P_{b2} = P_C = \frac{\lambda_t^{(C-C_h)} \lambda_h^{C_h}}{C! \mu^C} P_0 \qquad [5.41]$$

The probabilities of a lost call and a failed call, P_{drp} and P_{ns}, respectively, are given by:

$$P_{drp} = \frac{P_{h1}(1 + P_{h2}P_{b2})P_{b2}}{1-(1-P_{b2})P_{h2}} \qquad P_{ns} = P_{b1} + (1-P_{b1})P_{drp} \qquad [5.42]$$

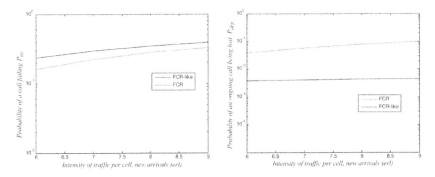

Figure 5.5. *Theoretical comparison between the performances of FCR and FCR-like strategies according to P_{ns} and P_{drp} IRIDIUM case, S = 10*

Figure 5.5 shows the analytical results for FCR and FCR-*like*. According to the probability P_{drp}, it can be seen that FCR-*like* performs well. Indeed, the probability of an ongoing call failing is higher for FCR as the number of guard channels becomes insufficient for the handover requests generated. This number grows while traffic increases, thus increasing the probability of handover requests failing. Whereas for FCR-*like*, *to ensure* success of the first handover implies a higher probability of handover requests failing and as a result a better probability of ongoing calls failing.

However, it can be seen that FCR has a higher probability P_{ns}, which is essentially due to the *locking-channel* mechanism used in FCR-*like*, which blocks a channel in its first transit cell from the start of the call in its source cell. Though this procedure improves the probability of ongoing calls being lost given that the success of the first handover is guaranteed, it does, however, lead to an increase in the probability of new calls being blocked, since fewer channels are available for them. This increase results in an increase in the probability of calls failing.

6

The Rescuing System

6.1. Introduction

Spotbeam handover is a major problem for mobile satellite networks; requests to switch from one cell to another are very frequent due to the relatively small size of the cells and the relatively fast speed of the satellites. The limited number of radio channels often causes the handover procedure to fail and ongoing calls are thus lost prematurely, resulting in poor network performance [ZHA 89, CIM 94, TSU 04].

In this research, we will look at another aspect of the handover problem which, to our knowledge, has not been dealt with before: calls that are lost shortly after they are initialized due to a failure in the handover procedure [MAC 79, RAY 91]. Given that these procedures take place frequently and the cells in low earth orbit (LEO) satellite systems are seen as small in relation to the mobility of the system, the number of these calls is high and the annoyance they cause is great.

As a solution to this problem, we will present a decision system based on the concept of fuzzy logic that rescues calls that last for just a few seconds before they are faced with a premature end, at the expense of calls that last for a relatively long time.

6.2. Fuzzy logic

Fuzzy logic is the result of *Lotfi A. Zadeh*'s research in 1965. A professor at the University of California, Berkley [ZAD 64], *Lotfi A. Zadeh*, is internationally renowned for his work on automation and systems theory, which tested the need to formalize the representation and the processing of imprecise or approximate knowledge, in order to process very complex systems involving, for instance, human factors [CHA 01]. Fuzzy logic is used to deal with imperfect knowledge [BOU 95].

6.2.1. *Definition of fuzzy subsets*

The concept of fuzzy subsets was introduced to avoid the sudden transition from one class to another (from the black class to the white class for example), to enable elements not to belong completely to one or the other (to be gray for example) or to partially belong to each (with a strong degree of the black class and a weak degree of the white class, in the case of dark gray). The definition of a fuzzy subset is a response to the need to represent imprecise knowledge; knowledge is imprecise either because it is expressed in natural language or it has been obtained with observation tools that produce measurement errors.

A fuzzy subset A of X is defined by a function of belonging that associates with each element x of X, the degree $f_A(x)$ between 0 and 1 with which x belongs to A:

$$f_A : X \to [0,1]$$

In the particular case where f_A only takes values equal to 0 or 1, the subset A is a classical subset of X. A classical subset is thus a particular case of fuzzy subset.

6.2.2. *Decisions in the fuzzy environment*

All aspects of decision-aid and decision-making have been studied in the fuzzy context to process situations where some knowledge leading to the decision is incomplete.

Decision-making can be based on fuzzy constraints *"the duration of the repairs will be approximately 120 h"*, on criteria whose characterizations are fuzzy *"comfort is satisfying"*, the objective of the decision-making can itself be fuzzy *"financial income must be much larger than last year's"*. Decision-making can be regarded as looking for, out of decisions of a set D, the most compatible decision with the values observed from the criteria of a set k.

The simplest form of decision in the fuzzy environment [BOU 95] consists of making a fuzzy objective G and a fuzzy constraint c, two fuzzy subsets of D. For instance, if D is the set of trains to *Alger*, the objective of the decision could be to choose a train that arrives *"relatively late"*, given the constraint stating that the train must arrive *"well in advance of the conference"*. A *fuzzy decision* is therefore the fuzzy subset d of D defined as $d = G \cap c$. The intersection $\cap$ can be classically associated with the minimum operator. An *optimal decision* is therefore obtained in the set of the elements of D where the function of belonging f_d of d reaches its maximum.

6.3. The problem

In LEO mobile satellite system (MSS) networks, a high number of ongoing calls are forced to end prematurely a short time after they initialize due to a failure in the handover procedure.

In the mobility model, we are assuming the time required for an mobile user (MU) to travel the maximum distance in a cell before initializing a handover request (i.e. the maximum stay time) t_{Msj} is given by [4.13]. If an ongoing call waiting for a channel in its transit cell to be freed reaches the limit of the overlap zone without being switched, the communication is lost. Therefore, the maximum period t_{MP} required for a call to travel from one end of the cell to another is equal to the sum of the maximum stay time and the maximum queuing time t_{qmax} [4.11] $t_{MP} = t_{Msj} + t_{qmax}$. Figure 6.1 represents the variation of this time period according to z.

It can be seen that in certain regions, this time period is very short. For instance, for 199.5 km < $|z|$ < 212.5 km, the values of t_{MP} are lower than 20 sec, which represents $T_m/9$. T_m is the average duration of communications and is generally supposed to be equal to 180 sec.

As a result, calls generated in these regions will probably need to be switched a short while after they are initialized. They thereby risk being forced to end prematurely if no channel is free in the transit cell.

The role of our decision system is to rescue calls facing a failed handover procedure a relatively short time after they are initialized.

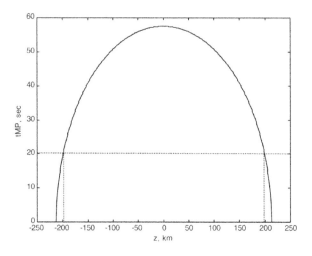

Figure 6.1. *Maximum period that a call can spend in a cell t_{MP} according to z*

To rescue these calls, the decision system we are proposing favors service to calls that have only lasted a "short" amount of time, rather than those that are in the transit cell of the latter and that have lasted a "relatively long" amount of time.

Presented in this manner, the problem seems complicated. Indeed, our reasoning is based on estimating "*short duration*" and "*relatively long*" and so making an adequate decision can seem difficult. However, in the presence of a decision-aid system, fuzzy logic can be seen as an important tool to add to the panoply of tools that are already available. Fuzzy logic interests us as there is great variability in the system and a lack of knowledge about the sum of possible situations.

6.4. Rescuing system [DEL 99]

We denote by CHF a call facing a handover failure and by CLD the oldest call (the call with the longest duration) in the transit cell of CHF.

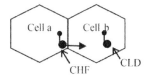

Figure 6.2. *A call facing its handover request being rejected*

The rescuing system (RS) is activated the moment a handover request is rejected. It compares the duration of CHF and CLD: if the duration of CHF is considered relatively short in relation to the duration of CLD, the latter is interrupted and its channel is used to rescue CHF.

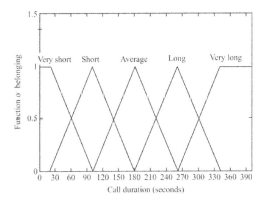

Figure 6.3. *Function of belonging of call durations*

To represent linguistic uncertainties (short duration, long duration, etc.) by numbers in the interval [0, 1], a function of belonging of five fuzzy subsets {very short, short, average, long, very long} is defined as follows.

The system's decision is defined by two fuzzy subsets {Rescue, Do Not Rescue}. Out of 25 possible cases, the system decides to make a rescue operation in the following 10 cases.

	CHF		CLD
IF	Very Courte	AND	Short
IF	Very Courte	AND	Average
IF	Very Courte	AND	Long
IF	Very Courte	AND	Very
IF	Short	AND	Average
IF	Short	AND	Long
IF	Short	AND	Very
IF	Average	AND	Long
IF	Average	AND	Very
IF	Long	AND	Very

Table 6.1. *Cases in which the RS decides to make a rescue operation*

In all other cases, the system decides not to rescue the call facing a premature end.

The fuzzy AND is the minimum operator and the optimal decision is taken equal to the maximum value in the fuzzy subset of the system's decision.

EXAMPLE.– To better explain the behavior of this system, let us look at a number of examples. The following table shows the degrees of belonging of different call durations in the five fuzzy subsets:

– if, for example, the durations of CHF and CLD are 20 and 200 sec, respectively, we thus have CHF (very short) and CLD (average). The decision is therefore rescue;

– if, for example, the durations of CHF and CLD are 80 and 100 sec, respectively, we thus have CHF (short) and CLD (short). The decision is therefore do not rescue.

Call duration (sec)	Very short	Short	Average	Long	Very long	Max
100	0	1	0	0	0	1 in S
20	1	0	0	0	0	1 in VS
150	0	0.3750	0.6250	0	0	0.625 in A
80	0.2500	0.7500	0	0	0	0.750 in S
200	0	0	0.7500	0.2500	0	0.75 in A
360	0	0	0	0	1	1 in VL

Table 6.2. *Degrees of belonging of different call durations*

7

Results and Simulation

7.1. Introduction

To study the dynamic case, we developed a simulation of our network, so we could evaluate its performance in different handover request management strategies and different channel allocation techniques [ZHA 89].

The simulation was developed in the *Matlab* language which is well suited to this study as its library has a number of distribution functions that are required for our application. We opted for a very long simulation time of 20,000 s, which is double the simulation time that tends to be used in the literature [DEL 96]. This guarantees the stability of the system and convergent results.

Through the intermediary of different Quality of Service (QoS) parameters, the simulation enables us to evaluate the handover request management strategies we have proposed and to compare them to other strategies that have already been used in the literature [LIN 95, SHE 98, WON 00, TSE 07, SAN 08].

This section will contain a study on the different aspects of the performance of the mobile satellite system (MSS) network. We will first deal with the problem of channel allocation and will, through the intermediary of probability results, show the effectiveness of dynamic allocation over fixed allocation. We will then discuss the effect of prioritizing service (giving priority) to handover requests over new call attempts. Subsequently, we will present the results concerning the two strategies we have proposed, Pseudo LUI (PLUI) and dynamic channel reservation (DCR)-*like* and will compare them to other strategies. Finally, we will present the results relating to the use of the rescuing system (RS) decision system.

Before this, we will present a simulated *folded network* that enables a *three-dimensional* simulation of the satellite network.

7.2. The (folded) simulated network

The simulated network is a parallelogram. In the literature, this has been a popular choice for studying cellular land systems. However, to adapt this topology to the characteristics of universal MSSs in which the cellular network has three dimensions, we will define a closed cell network to obtain a complete belt of interfered cells for each cell [TEK 92].

We denote by N, the number of cells per side in the simulated cellular network. Each cell in the network has an identification number: $n = 1, 2, 3, ...N$, (see Figure 7.1, for $N = 7$). The set of cells belonging to the simulated network is indicated by $P = \{1, 2, 3,..., N^2\}$. What's more, an oblique reference originating from the center of cell no. 1 is considered.

If we normalize the distance between the centers of two adjacent cells, each cell center is indicated in this reference by two whole numbers (ζ, η). Let us suppose that the interference cell belt of cell x, $I(x)$, is formed by two rows of adjacent cells. Therefore, the two outermost rows of cells in the parallelogram network have an incomplete belt of interference cells and form the set B of edge cells. Consequently, the cells belonging to the set $P - B$ are called the "central" cells.

Let us consider a specific cell $z \equiv (\zeta_z, \eta_z)$ on the border of the network. A specific cell $k \in I(z)$ has the following coordinates: $k \equiv (\zeta_z + a, \eta_z + b)$, where a and $b \in \{0,1,2-1,-2\}$, $a+b \in \{-2,-1,0,1,2\}$ and $|a|+|b| \neq 0$.

If $k \in P$, the interference cell exists.

However, if $k \notin P$, the interference cell does not exist; therefore k becomes a *false* interference cell of z, resulting from the following rule: the coordinates of the interference cell $k \equiv (\zeta_z + a, \eta_z + b)$ are transformed into $k \equiv (\zeta_k, \eta_k)$ with $\zeta_k = (\zeta_z + a) \mod (N)$ and $\eta_k = (\eta_z + b) \mod (N)$. As a consequence, a cell on a border of a simulated network interferes with the border cells on the other side of the network. Figure 7.1 shows the *false* interference cells for border cell no. 48.

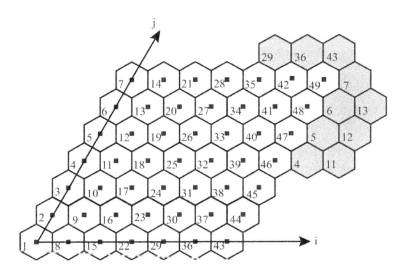

Figure 7.1. *Folded network with false interference cells*

When a closed network is used, an mobile user (MU) with an ongoing call who leaves one side of the network enters by the other side, thereby perfectly reproducing the characteristics of an infinite network. This is a considerable advantage as the performance of a resource management technique is being jointly evaluated with a user mobility model. Due to the chosen topology of the simulated network and the size of the interference cell belt, if we use the *cost-function* defined in [3.3], the previous folded network is only valid for values of N that are whole multiples of 7. Only for these special values of N do the *false* interference cells around a border cell have a nominal channel allocation according to the regular fixed channel allocation (FCA) model $F_D(z)$, used for the network.

7.3. Simulation results

To evaluate the strategies we proposed in the previous chapter, we will now develop a simulation using the programming language Matlab. Our study is subdivided into three parts: a part concerning the *Queuing Handover* (QH) strategies first in first out (FIFO), last useful instant (LUI) and PLUI, a part concerning the guaranteed handover strategies DCR and DCR-*like* and a part concerning the RS.

Part 1: QH strategies

On the basis of the model presented in Chapter 3, we have developed a simulation with the following suppositions:

– the size of the folded network used in the simulation (following the most popular choice in the literature) is 7 cells per side, therefore a total of 49 cells [HON 86];

– the average duration of calls is $T_m = 180$ s;

– the interference cell belt is formed by two rows of cells;

– the number of channels available in the system is 70;

– the new call arrival process is an independent Poisson process from cell to cell with an average call arrival rate per cell equal to λ;

– the capacity of the queue is assumed to be infinite.

7.3.1. *Verifying the simulation: a comparison with the analytical results of the FCA-QH case with different queuing strategies*

To verify our simulation, we first used it for the FCA-QH case with different strategies before comparing the results we obtained with the analytical predictions presented in the previous section. This comparison is presented in [ZHA 89].

Though a good concordance between the two was noted, a small difference was observed. This difference is essentially due to the supposed simplifications in the analysis concerning the *pdf* of the maximum queuing time, the *pdf* of the channel holding time and the arrival process of handover requests.

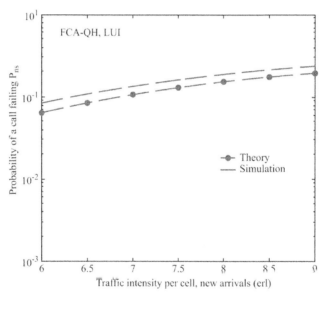

a)

Figure 7.2. *a) FCA-QH with LUI queuing strategy, comparison between the simulation results and the analytical prediction (IRIDIUM case, S = 10)*

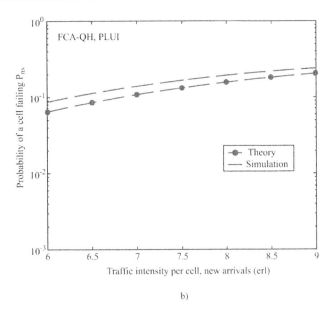

Figure 7.2. *b) FCA-QH with PLUI queuing strategy, comparison between the simulation results and the analytical prediction (IRIDIUM case, S = 10)*

Before presenting the evaluation of our PLUI strategy, we would like to illustrate the advantages of dynamic channel allocation and the advantages of the use of strategies that give priority of service to handover requests.

The performance evaluation results are probabilities, new call rejections P_{b1}, failed handover procedures P_{b2}, the interruption of an ongoing call P_{drp} and call failure P_{ns}.

The QoS parameters that are directly experienced by MUs are P_{b1} and P_{drp}. With regards to these parameters, the E.771 norms outlined by the International Telecommunication Union (ITU)-T for land cellular systems are $P_{b1} \leq 10^{-2}$ $P_{drp} \leq 5\ 10^{-4}$ [ROS 95].

Although strict for Low Earth Orbit (LEO) MSSs, these norms have been adopted in the literature for these systems [DEL 99].

In reference to the IRIDIUM case, we have $n_h < 5$ handover/call [3.32]. Therefore, from [3.33] and to meet the ITU-T norms relating to P_{drp}, P_{b2} must be $\leq 10^{-4}$.

7.3.2. A comparison of FCA and DCA, DCA-QH & FCA-QH simulation using LUI

Figure 7.3 shows the effectiveness of dynamic channel allocation by comparing it to fixed channel allocation for the QH case with the LUI strategy.

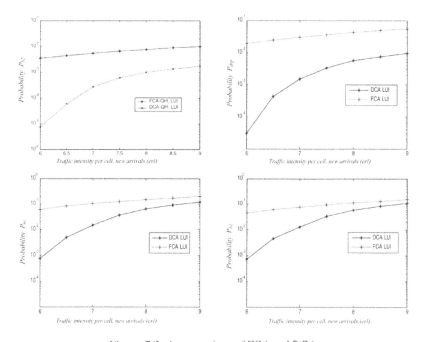

Figure 7.3. *A comparison of FCA and DCA*

It can be seen that DCA performs better than FCA. Indeed, it can be noted that the probability of new call rejections P_{b1} considerably improves with dynamic channel allocation. This is also the case for the probability of a failed handover procedure P_{h2}. We note that the most critical demand, i.e. the demand that principally limits the capacity of the system, is P_{h2}. The FCA technique does not meet any of the requirements of P_{h1} and P_{h2} in the traffic intensity considered.

This is because the DCA technique uses channels efficiently: the network has the option to use any channel in any cell while paying heed to a number of interference constraints. DCA adapts to variations in network traffic and therefore improves QoS by using network resources more efficiently.

7.3.3. *A comparison of NPS and QH, DCA-NPS & DCA-QH simulation*

To demonstrate the effect that using a strategy that prioritizes handover requests over new call attempts has on the performance of an MSS network, we will present the performance of a network that does not use any of these strategies, a case known in the literature as the *Non Prioritization Scheme* (NPS). A comparison of DCA-NPS and DCA-QH with LUI is presented in Figure 7.4.

The difference is remarkable, whether it be for the probability of call rejection or the probability of failed handover request. Indeed, for the probability of call rejection, the DCA-NPS case performs very well in relation to DCA-QH LUI, whereas for the probability of failed handover, DCA-QH LUI considerably improves this parameter.

Therefore, in relation to the NPS system, the QH strategy (regardless of the queuing discipline and the channel allocation technique used) enables P_{h2} to be reduced considerably at the expense of an increased value of P_{h1}.

Indeed, using a queuing strategy favors the service of handover requests over the service of new calls. This causes an increase in the probability of new call rejection, but considerably improves the probability of losing ongoing calls. The increase of P_{h1} is seen as tolerable from the user perspective; the loss of an ongoing call is less acceptable than the rejection of a new call.

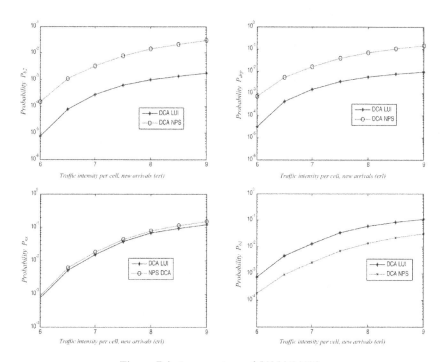

Figure 7.4. *A comparison of QH LUI-NPS*

7.3.4. Comparison of QH strategies, DCA-QH FIFO, LUI, PLUI simulation

The results obtained for the DCA-QH case are presented in Figure 7.5. First, it must be noted that the figure clearly shows that P_{b1} is independent of the queuing strategy adopted for handover requests, as has already been mentioned in the analytical part.

Indeed, for new call establishment requests, the concept of putting handover requests in a queue has no effect on them whatsoever, given that handover requests would be prioritized.

Conversely, for the probability of failed handover procedure, it can be noted – as was discussed in Chapter 5 – that (for the dynamic allocation case) the choice of a specific order of priority in the queue has an effect on the performance of the network.

Figure 7.5 presents the probability of failed handover requests. It can be noted that LUI performs better than FIFO and PLUI; however the latter enables a reduction in P_{b2} in relation to the FIFO strategy.

The performance obtained by FIFO is, therefore, the most mediocre. Indeed, the system in this strategy does not take into account any criteria concerning the urgency of one call over another; it simply ranks calls according to their order of arrival and serves the oldest request in the queue when a channel is freed in its transit cell.

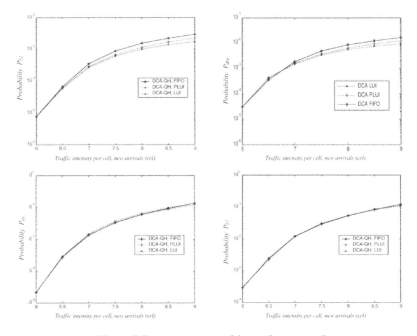

Figure 7.5. *A comparison of the performance of FIFO, LUI and PLUI (IRIDIUM case)*

According to the results, it can also be noted that the performance of PLUI is close to the performance of LUI. The reason for this is that the average number of handover requests generated per call attempt in the IRIDIUM case is about 5 handovers/call. The majority of queuing requests are generated from calls that have already been switched and that need to be switched again; they are therefore, in the PLUI strategy, put in a queue according to the LUI strategy.

As predicted, the problem of the central zone of cells only has a limited impact on the effectiveness of PLUI as the latter performs well. Without this problem, the difference between the performance of LUI and PLUI would have been smaller.

Part 2: Strategies using the concept of guard channels

The following considerations have been taken during the simulation of the DCR and DCR-*like* technique:

– the mobility model presented in Chapter 4 is used. The simulated network is composed of rectangular cells forming a coverage route on the surface of the earth (it is supposed that the calls are exclusively generated in the central zone of the cells);

– the arrival rate of new calls in each cell is λ;

– the average call duration is 180 s;

– FCA is used and each cell has $C = 10$ channels;

– a uniform traffic model is considered;

– a model of 7 cells is used; a call leaving the 7th cell initializes a handover request in the 1st cell (folded or closed network);

– the side of the cell R is equal to 425 km;

– the interference cell belt is formed by rows of cells.

7.3.5. *Verifying the simulation: a comparison with the analytical results of the FCR and FCR-like case*

As before, to verify our simulation, we first used it for the FCR and FCR-*like* cases and then compared them to the analytical results we obtained in the previous section. This comparison is presented in Figure 7.6.

A good concordance was noted between the analytical predictions and the simulation results. The small difference we observed is essentially due to the supposed simplifications in the analysis concerning the *pdf* of channel holding time and the arrival process of handover requests.

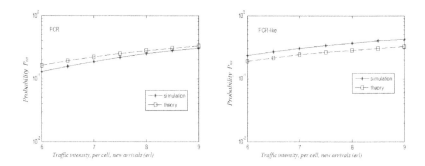

Figure 7.6. *FCR and FCR-like comparison between the simulation results and the analytical prediction (IRIDIUM case, C = 10)*

7.3.6. A comparison of DCR and DCR-like

Here we will consider DCR and DCR-*like* strategies without a QH strategy. In this case, the handover request is therefore initialized at the limit of the cell and the overlap zone is not taken into consideration.

The performance evaluation results for the probabilities, the new call rejection P_{b1}, the failed handover procedures P_{b2}, the loss of an ongoing call P_{drp} and the failed call P_{ns} are shown in Figure 7.7 for DCR and DCR-*like* techniques.

These results show a significant difference between the performance of DCR and DCR-*like*. Indeed, it can be seen that the DCR-*like* strategy has a better probability of a failed handover procedure in comparison to DCR. This is not only due to the guard channels that have been reserved for requests, but because a *locking channel* mechanism has been adopted for new calls that guarantees the success of the first handover request. Nevertheless, these reservations are the cause of an increase in the probability of new calls being blocked, since a new call is only accepted if there is a free channel in its source cell and in its transit cell. This reservation considerably reduces the number of free channels in relation to new call establishment requests, particularly when traffic intensity is high, and thus explains the poor performance of DCR-*like* according to P_{b1}.

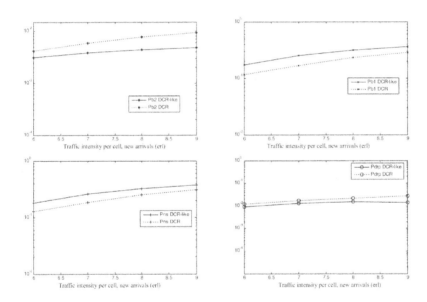

Figure 7.7. *Performance of DCR and DCR-like techniques*

The *locking channel* mechanism causes the large divergence between DCR and DCR-*like* results. The fact, channels are closed for the first handover request, affects the capacity of the system as fewer calls are admitted into the network. The locking channel mechanism also causes the low probability of failed handover requests because the success of the first request is guaranteed.

The aim of this research was to show that, using certain advantages of the LEO satellite network mobility model, it is possible to determine certain parameters that are required for handover request management when the exact position of the MU is unknown.

For this case, our idea was developed by reformulating LUI and DCR strategies into PLUI and DCR-*like* strategies. In this context, we consider that the results we obtained using these two reformulated strategies based on the EPM are satisfactory as they show that the latter is useable.

Nevertheless, the results obtained do not provide the same outcomes in terms of performance when compared to the original strategies. Through PLUI performs similarly to LUI, DCR-*like* gives different results to DCR. These differences are, of course, the consequence of the changes made to the original strategies in order to adapt them to new conditions: the absence of all information at the start of communications.

Part 3: the RS

To determine the effect of our RS on the behavior of the network, the duration of interrupted calls was collected during the simulation (DCA QH-LUI) for each value of traffic intensity in the two cases: With RS and Without RS. We adopted the same suppositions made in part 1 of the study.

The histograms presented in Figure 7.8 clearly show that the majority of calls in the Without RS case were lost with relatively short durations. Indeed, more than 60% of durations were less than T_m for different traffic intensity values. It can also be seen that a large number of calls were lost with very short durations. This number increases considerably with an increase in the traffic: indeed for $\lambda = 7$erl, we have 8% of lost calls with durations less than 30 s. This percentage reaches 20% for $\lambda = 10$erl.

When using the RS, the results displayed in Figure 7.9 show that all the durations of lost calls for different values of traffic intensity are higher than 60 s and that the number of lost calls with durations higher than 180 s increased considerably. This demonstrates a significant increase in the average duration of lost calls. Figure 7.10 shows the variation of the average duration of lost calls according to the traffic intensity in both cases, With RS and Without RS. For the former case, the average duration of lost calls is reduced by almost 190 s to about 140 s when there is an increase in traffic intensity, whereas in the latter case, it exceeds 400 s for the different values of traffic intensity.

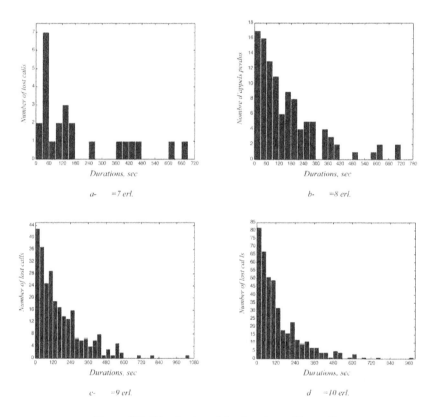

Figure 7.8. *Distribution of the durations of lost calls for different values of λ without RS*

In this case, therefore, the RS guarantees at least 60 s of continual communication for every call initialized, without fear that it will suddenly cut out, and thereby considerably improves the average value of the durations of lost calls.

The RS simultaneously forces a handover request to succeed and an ongoing call to be broken. Therefore, in terms of probability, the system does not affect the performance of the network (Figure 7.11), due to the fact that even if the call is rescued, the rejected handover requests will still cause the loss of an ongoing call.

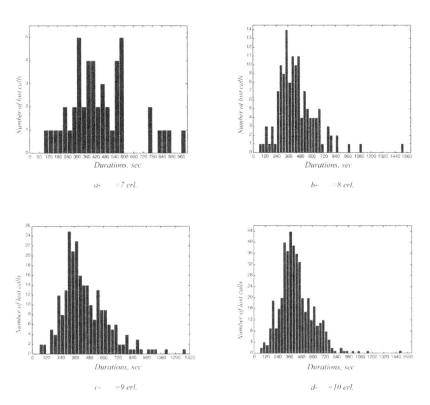

Figure 7.9. *Distributions of the durations of lost calls for different values of λ with RS*

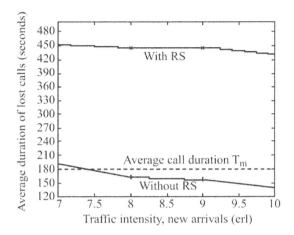

Figure 7.10. *Average durations of lost calls*

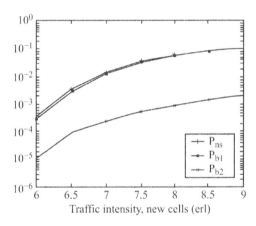

Figure 7.11. *Performance of the DCA QH-LUI network with RS*

8

PAB for IP Traffic in Satellite Networks

8.1. Introduction

Satellite networks come off badly from a comparison with land networks in terms of Quality of Service (QoS): they have a long propagation delay, a more limited bandwidth and more frequent transmission errors [DUR 01, KOT 01, KOT 00]. Delays, the bandwidth and the error rate must therefore be very carefully managed so as to satisfy users. QoS management is therefore a major concern in satellite networks than it is in land networks [HAS 04, IBN 04, EVA 05]. This chapter will present the new DiffServ rule based on the property of bandwidth allocation during congestion, which is known as the proportional allocation of bandwidth (PAB) [ETK 07, XIE 04]. PAB can be used for geostationary earth orbit (GEO), medium earth orbit (MEO) and low earth orbit (LEO) satellite networks when all flows are congested. PAB makes part of the bandwidth available in proportion to the subscribed information rate. Nevertheless, in the current Internet system (including the satellite part), users can have different service requirements. The most important parameter, the committed information rate (CIR), varies a great deal between one user and another [KIM 03]. A user with a higher CIR generally pays more than a user with a lower CIR. Consequently, during congestion, a user with a higher CIR expects to be attributed more bandwidth than a user with a lower CIR [KIM 03]. We will define a new method for allocating bandwidth called PAB. In this method, the bandwidth must be allocated in proportion to the CIR of competing flows. The effectiveness of the satellite network depends on the QoS offered to clients. We will avoid flow state information being stored by encoding the ratio of flow data to its CIR, in the form of a label on its packets. When returning to the core of the network, the routers use these labels to differentiate packets during congestion. All labeling is done at the source, where the first ingress element in the satellite network, which has information about the source CIR, is located. The satellite router (SR) processes the packets according to their functions.

The results from the simulation show that our bandwidth allocation technique using CIR performs very well and allocates bandwidth proportionally by storing state information in the configuration of different satellites.

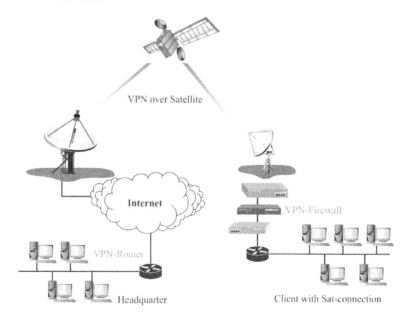

Figure 8.1. *General overview of a satellite network*

8.2. Proportional allocation of bandwidth

This principle is illustrated in Figure 8.2 according to the definition of max–min fairness in [STO 98]. The total bandwidth (left-hand column) is divided as follows:

– the flows wanting to carry less traffic than the equal sharing are satisfied (source A);

– the excess resources available are shared equally between all the sources wanting to benefit from a bandwidth that is higher than the value of the fair sharing. As such, source A enables B and C to benefit from additional throughput that is divided fairly between them.

– sources that are not satisfied have the same capacity allocation.

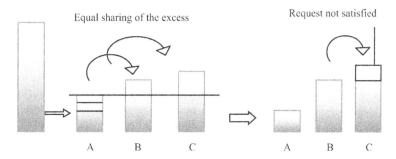

Figure 8.2. *The principal of max–min fairness [HUA 01]*

We can therefore allocate a bandwidth to each flow by:

$$Alloc\ (i) = Min\{send(i),\ rr\} \quad [8.1]$$

$$\sum Alloc\ (i) < Available\ bandwidth \quad [8.2]$$

Where send(i) is the throughput of the flow ith and **rr** is the maximum rate that satisfies the unfairness [8.2]. Any flow sending more than **rr** will have its throughput reduced to **rr**. Therefore, the underlying principle of PAB is that the allocation of bandwidth must be proportional to the CIR of the flows sharing the link. The CIR of a flow is one of the most important parameters of service for a flow. It is therefore important to consider both the throughput of data flow and its CIR to allocate bandwidth. As a result, in PAB, the allocation of bandwidth is given by:

$$\text{Alloc}(i) = \text{Min}\{\text{send}(i), \text{frac} * \text{CIR}(i)\} \qquad [8.3]$$

$$\sum \text{Alloc}(i) < \text{Available bandwidth} \qquad [8.4]$$

Here, CIR(i) is the CIR of the flow ith and frac is the maximum of the multiplying fraction (between 0 and 1) that satisfies the above unfairness. The frac determines the maximum throughput of a flow as a fraction of its CIR.

If the throughput of a flow is below what is permitted frac* CIR, then it does not suffer any packet loss. Moreover, if the flow has a throughput lower than its authorized CIR fraction, the excess of the bandwidth is shared between the other flows in proportion to their CIR. No flow is authorized to send more than its CIR during congestion. The throughput of any sent flow is reduced to its maximum authorized throughput. Therefore, PAB differentiates between the flows and allocates a bandwidth proportional to the CIR of the flows.

8.2.1. *Implementation of PAB*

Our technique for implementing PAB is based on the principles of differentiated services [ZHA 95b, EIN 01, PAR 04, DUR 02]. To avoid the state flow information in the core router, our technique uses labels to show the ratio between the flow rate (FR) and the CIR of the flow. The packets are marked with labels at the ingress of the satellite network. The ratio between the FR and the CIR of each flow is encoded as a label on the flow packets. In the center of the satellite network, bandwidth is allocated using these labels to differentiate between the packets.

Our technique for implementing PAB involves two main components: labeling packets returning from the ingress of the network and dropping these packets at the router in the core of the network.

8.2.1.1. *Packet labeling methodology [RAY 03, BAO 05, NIE 06]*

Packets are labeled at the source or at the ingress router. The ingress router has knowledge of CIRs of all the sources that are connected to it. The labeling mechanism marks the packet flow with different labels depending on the ratio between the FR to the CIR. The total number of labels is fixed for all flows, but the number of label values used at any time for a source depends on the FR. As the ratio between the FR and the CIR increases for a flow, more packets will be marked as low priority. We will describe how this mechanism marks packets according to the throughput and the CIR of the flow. This mechanism marks the packets of two flows with the same FR but with different CIRs. The flow with a low CIR has more packets with a low priority than a flow with a higher CIR, as shown in Figure 8.4. A label is not associated with a particular rate numerically. Each label is associated with just one fraction between 0 and 1. In Figures 8.3 and 8.4, we assume that there are just four labels and all labels are associated with the same fraction value 1/4.

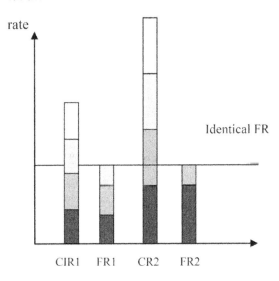

Figure 8.3. *Packet labeling with the same FR and different CIR*

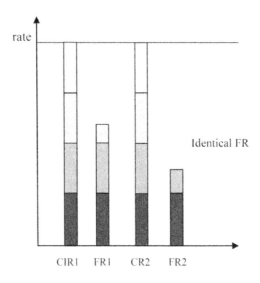

Figure 8.4. *Packet labeling with the same CIR and a different FR*

To label the packets, several token buckets are used at the source or at the ingress router, which knows the CIR of the source. The source must be labeled and it sends data equal to or lower than its CIR, but its average data throughput should never exceed its CIR. So, the sum of the token rates of all the token buckets must be equal to the CIR of the source. The CIR is thus divided between all the token rates. Consequently, the token rate of one token bucket is a fraction of the source's CIR. This fraction is equal to the fraction associated with the label corresponding to this token bucket. The sum of all the fractions associated with the labels is 1. Therefore, the sum of the bucket rate is equal to the CIR of the source. The throughput determines the actual label values the packets will obtain. As the ratio of throughput to CIR increases, more labels will be marked as low priority. For a flow to comply with a token bucket, it is necessary that all its packets have enough tokens in the queue and that the FR never exceeds its CIR.

Figures 8.5 and 8.6 show a pictorial representation of the token system used to label packets. In Figure 8.5, the packet (I) waits to remove the tokens from a token bucket. Buckets 1 to $K-1$ do not have enough tokens required by the packet (I). Therefore, the packet (I) removes tokens from bucket K. Tokens in the token bucket K have decreased due to the consumption of tokens by packet (I), as shown in Figure 8.6. The packet (I) removes tokens from bucket K, and packet (I) is marked with label value K.

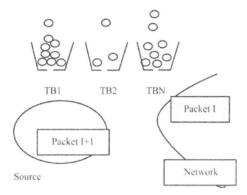

Figure 8.5. *After packet labeling*

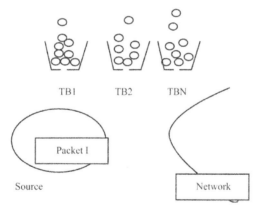

Figure 8.6. *First packet labeled*

8.2.1.2. *Packet dropping mechanism at the core routers*

At the satellite core router, in the event of congestion, packets are dropped depending on their labels [EKI 02, CHE 05, AKY 02, AWE 00, KUS 14]. The formula for calculating the average size of the queue is the exponential average technique [MIS 01a]. Following changes to the average length of the queue, packets are dropped due to a change in probabilities. The drop probability for a packet with a low priority label is higher than for a packet with a high priority label for the same queue size. The queue management mechanism used in our technique is multilevel-based queuing (MLBQ) [NIC 03, OUE 07].

MLBQ is designed to differentiate between several levels of priority in the core router and the drop probability for a packet is calculated from its label and the average size of the queue. One average queue size is maintained for all packets. There are n label values and n sets ($\min_{th}$, $\max_{th}$, P_{max}) exist in the core router. When $\min_{th}$ indicates the minimum threshold, $\max_{th}$ indicates the maximum threshold and P_{max} indicates the maximum drop probability.

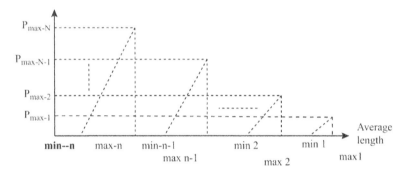

Figure 8.7. *Drop probability for packets in the core router*

When a packet labeled with a value k arrives in the router, the $\min_{th-k}$ and $\max_{th-k}$ P_{max-k} are used to determine whether this packet should be dropped. The priority of the different labels is obtained by choosing values for the thresholds and the drop probability. These values are indicated in Figure 8.7. A label with a low priority is given thresholds that have a high drop probability, even when the average queue length is low, whereas a label with a high priority is given thresholds that have a low drop probability, even when the average queue length is high. The highest priority is label 1 and the lowest priority is label N. Labels with a low priority have a high maximum drop probability.

8.3. Determination of the label fraction

We have studied three different sets of fraction values that can be attributed to the labels. These three sets of fraction values have specific properties.

8.3.1. *Equal fractions*

All fractions are of equal values. Therefore, if there are N labels, then each label has a fraction value of $1/N$. The sum of the fractions is 1.

8.3.2. *AP fractions*

The fractions form an arithmetic progression (AP). Unlike equal fractions (EFs), the values for AP fractions are not identical. To achieve better granularity while providing PAB, the smallest values in AP are associated with the highest priority among the labels. So, the fractions will have values given by:

$$a, a+d, a+2d, a+3d, \ldots, a+(N-1)d \quad [8.5]$$

For simplification, we assume that "a" is equal to "d". Therefore, this gives the values of the fractions as:

$$d, 2d, 3d, \ldots, Nd \quad [8.6]$$

Given that the sum of the values of the AP fraction must be 1, we obtain the following value for "d":

$$D = 2/N*(N+1) \quad [8.7]$$

For eight labels, the values of fractions are: 1/36, 2/36, 3/36, 4/36, 5/36, 6/36, 7/36 and 8/36.

8.3.3. *GP fractions*

The fractions can also form a geometric progression (GP). Similar to AP, the values of the fractions are assigned so that the priority labels have smaller values in the GP. Therefore, the fractions are given by:

$$a, ar, ar^2, \ldots, ar^{(N-1)} \qquad [8.8]$$

Given that the sum of fractions must be 1, the value of the sum is set to 1:

$$sum = \frac{a(1-r^N)}{1-r} = 1 \qquad [8.9]$$

As in AP, to simplify calculations, we assume "a" is equal to "r". As such, the values of the fractions are given by:

$$r, r^2, r^3, \ldots, r^N \qquad [8.10]$$

The value of r is given by the equation below when the sum of the fractions is unity:

$$r(2-r^N) = 1 \qquad [8.11]$$

For $N = 8$, r takes a value approximately equal to ½. Therefore, the fractions have the following values: 1/2, 1/4, 1/8, 1/16, 1/32, 1/64, 1/128 and 1/128. The last two fractions are equal such that the sum of the fractions is 1.

Figure 8.8 shows the values of the fractions attributed to labels for the three types of fractions: EF, AP and GP. With the same CIR and the same FR, the number of high priority packets is higher in EF than it is in AP and GP fractions. However, for the same CIR and the same FR, the number of low priority packets is lower in EFs than in AP fractions, which is lower than in GP fractions.

8.4. Simulation and results

In our simulations, we compared the performance of our implementation technique (PAB) with that of the random early discard (RED) technique [ROS 99, MIS 03b, MIS 02, ZEN 07, MAL 10, FER 11, CHE 12]. The performance of PAB in single congested links for satellite networks was studied. Likewise, the performance of PAB in multiple congested links in LEO and MEO satellite networks was also studied.

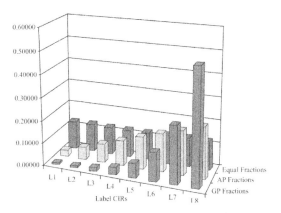

Figure 8.8. *Values of Equal, AP and GP fractions*

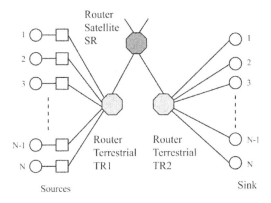

Figure 8.9. *Single congested link in satellite networks*

8.4.1. *Single congested link [CAO 09, TAL 06, VAL 07, AKY 01a, SEL 06]*

We used the NS-2 simulator to implement the DiffServ domain in the network. It is necessary to differentiate between edge routers and core routers. The packets are marked with labels at the ingress of the networks in the ingress routers, which are the gateways to the satellite network. SRs use the labels to provide services and drop packets during congestion. When one link is congested, the network configuration is shown in Figure 8.7. There are N flows sharing a single congested link, which is known as a bottleneck link [MIS 02]. The bottleneck link is the satellite link between the terrestrial router TR1 and the terrestrial router TR2 through the SR. The same results are obtained when a SR-TR2 is congested.

The CIR of the ith flow was set at (i/N *500) Kbps. The number of flows sharing the link varies from 3 to 32. The capacity of the bottleneck link is 1 Mbps. The link delay for GEO was set at 125 ms per route, 25 ms for LEO and 50 ms for MEO. The capacity of the link queue was 100 packets. The packet size of Transmission Control Protocol (TCP) flows was set at 1,000 bytes. The packet size of CIR flows was set at 210 bytes. The parameters for the eight levels of thresholds at the core router are indicated in Table 8.1.

For the token buckets at the ingress routers, the bucket size was set at 80,000 bytes. The token rate for the bucket of each label is determined by the CIR fraction associated with this flow label.

By definition of PAB, each flow must get a share of bandwidth, which is proportionate to its CIR. The measure we used to calculate the effectiveness of the PAB is obtained as follows. The throughput ratio of the ith flow [TR(i)] is defined as the ratio of throughput of ith flow to the sum of throughputs of all flows going through the same link:

$$TR(i) = \frac{Throughput \ of \ flow}{\sum Throughput \ of \ all \ flows} \qquad [8.12]$$

The CIR ratio for the ith flow [SR(i)] is defined as the ratio of the CIR of the ith flow to the sum of the CIRs of all the flows going through the same link:

$$SR(i) = \frac{CIR \ of \ flow}{\sum CIR \ of \ all \ flows} \qquad [8.13]$$

The allocation ratio for the ith flow [AR(i)] is defined as the ratio of TR (i) to SR (i):

$$AR(i) = \frac{TR(i)}{SR(i)} \qquad [8.14]$$

The measure of the fairness index is defined as follows:

$$Proportionality\ index = \frac{\sum AR(i) * \sum AR(i)}{N * \sum [AR(i) * AR(i)]} \qquad [8.15]$$

The proportionality index is reversely proportional to the difference between the allocation ratios of the different sources. The allocation ratio of different sources differs according to the value of the index which decreases in proportionality.

The flows in the network are User Datagram Protocol (UDP) [GAR 03, GU 07] or TCP [MIS 03b, MIS 02, AKY 01a, GOY 99, JAI 99, HEN 99, XU 04, ZHA 04b, WAL 07]. Experiments have been carried out with three different combinations of flows. In the first experiment, all the flows are UDP. Each source sends a flow in the network and the flow reaches the destination at the other end of the network. Experiments were conducted with the number of flows varying from 3 to 32. The sources sent data randomly between 10% and 200% of their CIR. Experiments were carried out using PAB and RED. Figure 8.10 shows the experiment results. The proportionality index varies according to the number of flows knowing that the flows are UDP.

In the second experiment, the flows are TCP. Figure 8.11 shows the experiment results. The proportionality index varies with the number of flows and the flows are TCP. In the third sequence of experiments, the flows of sources are mixed TCP and UDP. The data throughput of UDP and TCP flows are the same as in the previous experiments. Figure 8.12 shows the experiment results. The graph shows the proportionality index according to the number of flows knowing that the flows are mixed.

Label	Maximum Threshold	Minimum Threshold	Drop Probability
1- High priority	80	90	1/50
2	70	80	1/45
3	60	70	1/40
4	50	60	1/35
5	40	50	1/30
6	30	40	1/25
7	20	30	1/20
8- Low priority	10	20	1/15

Table 8.1. *Parameter for core router*

8.4.1.1. *GEO satellites*

In the case of GEO satellite networks, the delay is 125 ms between the ground base and the satellite; therefore, the total delay between the two land stations is 150 ms. The flows in the network are either UDP or TCP. Experiments were conducted with three different combinations of flows, as shown in Figures 8.10–8.12. These figures are comparable with the principal figures in [TAL 05, OH 05, MAT 07, TAL 09].

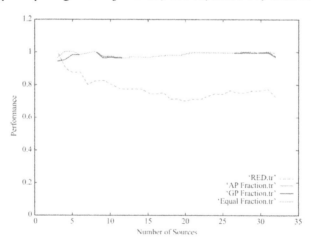

Figure 8.10. *Performance with UDP flows in GEO satellite networks*

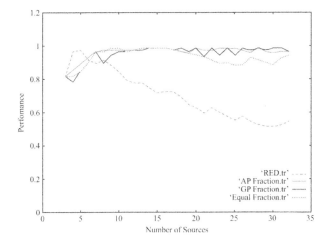

Figure 8.11. *Performance with TCP flows in GEO satellite networks*

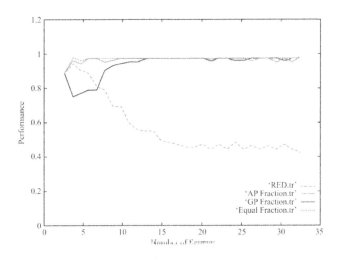

Figure 8.12. *Performance with UDP and TCP flows in GEO satellite networks*

In the first case, when the flows are UDP, it can be seen that the number of flows increases: this is a good performance for PAB. However, for RED, performance drastically drops as the number of flows increases. This is due to the fact that RED has no knowledge of the CIR and cannot tell the difference between the flows based on their CIR. Allocation of bandwidth by RED therefore does not follow the principles of PAB. RED was not, in fact, designed to enable sharing of resources proportional to CIR. The reason we compare PAB and RED is to show the advantages of the PAB service for users. In the second set of experiments, the sources are from TCP flows. TCP flows are sensitive to congestion. In this case, the TCP protocol of the flow tends to share the bandwidth fairly between the flows. Our technique enables us to obtain the sharing of proportional bandwidth using labels and thus performs well. In RED, the performance is considerably worse than the performance of UDP flows. During congestion, TCP flows reduce the rate of sending data so that the rate of all flows is equal. RED cannot distinguish between flows and thus performs badly.

Mixed flows, i.e. TCP and UDP, have throughput data that is comparable to the previous experiments. UDP is not sensitive to congestion and TCP is sensitive to congestion. Thus, UDP flows try to obtain all the bandwidth, whereas TCP flows get very little bandwidth. Our technique provides good protection of TCP flows from UDP flows and achieves excellent performance to offer a better differentiation. However, for RED, TCP flows are not protected and therefore RED does not provide enough protection of TCP from UDP and there is no delay guarantee.

The performances of the three fractions are almost identical. For MEO and LEO satellite networks, experiments are based on comparing REDs and for PAB, we use AP fractions for simplification purposes.

8.4.1.2. MEO satellites

In the case of MEO satellite networks, the delay is 50 ms between the ground base and the satellite, and therefore the total delay between the two land stations is 100 ms. The flows in the network are either UDP or TCP. Experiments have been conducted with three different combinations of flows in the same way for GEO satellites, as shown in Figures 8.13–8.15.

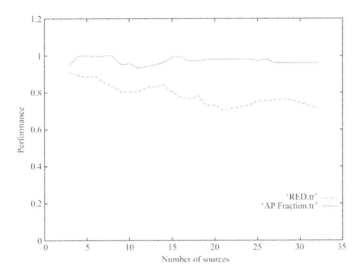

Figure 8.13. *Performance with UDP flows in MEO satellite networks*

In this case, the flows are UDP. It can be seen that when the number of flows increases, PAB performs well as it offers better differentiation. However, for RED, performance drops drastically as the number of flows increases. This is due to the fact that RED does not differentiate between the flows. UDP is not sensitive to congestion [HLU 93, AWE 04, PAT 05].

For TCP flows, the performance of PAB is still very good, although the TCP flow is sensitive to congestion. If there is congestion, the bandwidth is shared equally between the flows. In comparison to UDP, the performance of RED is very poor: during congestion, throughput is reduced until the flows receive their fair share. In this case, RED cannot guarantee that a distinction is made between the flows.

In the case of mixed flows, some sources send UDP flows and others send TCP flows. The performance of PAB is better than that of RED as TCP is sensitive to congestion, unlike UDP. So, UDP takes the maximum bandwidth and TCP takes the minimum for PAB, protecting TCP flows from UDP flows and thus giving better results.

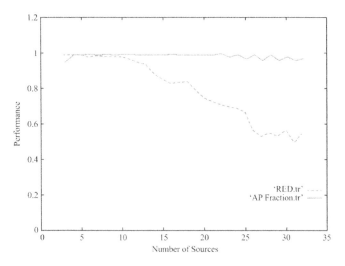

Figure 8.14. *Performance with UDP flows in MEO satellite networks*

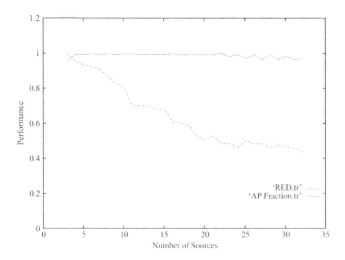

Figure 8.15. *Performance with TCP flows in MEO satellite networks*

8.4.1.3. *LEO satellites*

In the case of LEO satellite networks, the delay is 25 ms between the ground base and the satellite. Therefore, the total delay between two land stations is 50 ms. Flows in the network are both UDP and

TCP. Experiments have been carried out with three different combinations of flows, as shown in Figures 8.16–8.18.

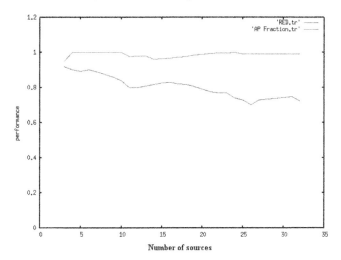

Figure 8.16. *Performance with UDP and TCP flows in LEO satellite networks*

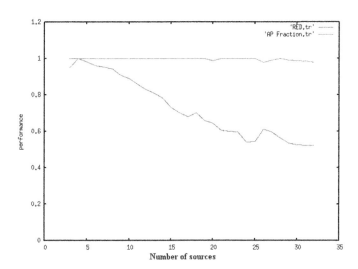

Figure 8.17. *Performance with UDP flows in LEO satellite networks*

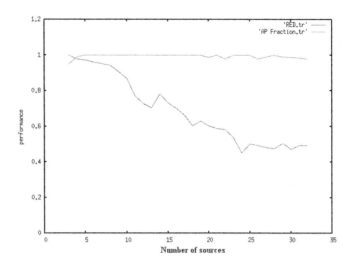

Figure 8.18. *Performance with TCP flows in LEO satellite networks*

The same result is obtained for LEO networks. From the graph, it can be seen that in the first case when the flows are UDP, the number of flows increases. This generates a good performance for PAB, but with RED, the index drops as the number of flows increases. The same can be said for TCP flow, which is good with PAB and very poor with RED. With mixed TCP and UDP flows, PAB provides very good results.

8.4.2. *Multiple congested link*

In the case of multiple congested links, the network topology for simulation is shown in Figure 8.19.

This configuration is identical to Figure 8.19; the flow travels different distances in the network. There are $N+1$ terrestrial routers. A terrestrial router is linked to the next terrestrial router via the SR. The satellite links connecting the routers have a bandwidth of 1.5 Mbps. The link delay for LEO satellite networks was set at 25 ms.

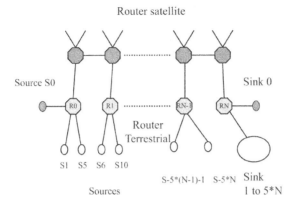

Figure 8.19. *Multiple congested links in satellite networks*

From router R0 to router RN−1, the flows enter the network and at router RN, all flows leave the network. In router R0, the flow S0 enters the network. At router Ri, the flows Si * 5 +1 to S(i +1) *5 enter the network. In each configuration of the test, the number of congested links varied from 2 to 5. Similar results were obtained using MEO constellations.

The performance of PAB in multiple congested links is defined as the ratio between the S0 flow and its CIR, divided by the ratio of the sum of the throughputs of all flows to the sum of CIR of all flows. This measure is the allocation ratio of S0 throughput.

$$Allocation-rate-AR(0) = \frac{\frac{Throughput(0)}{CIR(0)}}{\sum \frac{Throughput}{CIR}} \qquad [8.16]$$

Two tests were conducted using TCP and UDP protocols as a flow of S0 source. According to Figure 8.20, it is clear that the performance of RED is poor with the UDP flow of S0. Figure 8.21 shows that the performance of RED with the TCP flow of S0 is very poor and almost drastically zero. With PAB, as the number of congested links increases, there is a variation in the performance. This is due to the fact that our technique is only an approximation for the PAB and uses a limited number of levels of priority. Moreover, the behavior of TCP varies considerably through the intermediary of the link, which depends on a threshold and the fraction authorized by CIR. In the case of high-level congestion, GP fractions are better adapted due to their higher granularity for high priority labels.

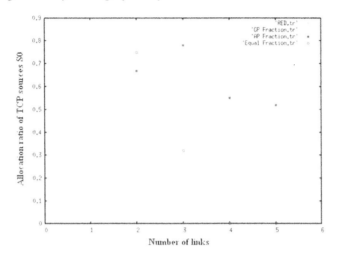

Figure 8.20. *Allocation ratio of UDP flows in MEO satellite networks*

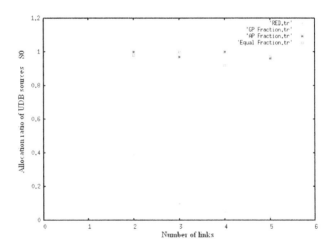

Figure 8.21. *Allocation ratio of TCP flows in LEO satellite networks*

8.5. Conclusion

This chapter has suggested the use of a new DiffServ rule based on the property of bandwidth allocation during congestion, which is called the PAB. PAB can be used for GEO, MEO and LEO satellite networks. During congestion, all flows can get a share of the available bandwidth that is proportionate to their subscribed information rate. We suggest applying PAB without recording the state flow in the center of the network, thus making the system scalable and simple. We showed via simulation the advantages of using PAB in Internet Protocol (IP) satellite networks, which also showed satellite networks dedicated to Internet traffic [AKY 01b, ZAH 02, VAR 03, ARA 05, ZAF 08, JAM 03].

General Conclusion

Given their various advantages (including a relatively low transmission power and a short transmission time), low earth orbit (LEO) satellites have become a subject of great interest for designers of telecommunication networks. Nevertheless, their short visibility in relation to an observer on the Earth creates a complex call management system. Indeed, the spotbeam handover procedure is considered a major problem due to its high frequency during a communication.

To manage channel allocation in these networks in the best way possible, a great deal of research has been presented in the literature to obtain a good Quality of Service (QoS). Some of this research presents good techniques providing good results in terms of the probability of new call rejection and loss of ongoing calls. Nevertheless, these techniques require a global positioning system to be integrated into the mobile satellite system (MSS). This integration is considered complex and makes designing the network more difficult.

First, and to resolve this problem, this study presented a method using the advantages of the mobility model of LEO satellite constellations and thus enabling different parameters about MUs with ongoing calls to be determined. These parameters are useful for handover request management techniques. The method uses the predictable movement of MUs throughout the network and the regular cell topology.

Indeed, the delay between two consecutive handover requests is the maximum stay time of the call in the cell. Using a *timer*, this amount of time can be estimated for each MU with an ongoing call, while the latter starts their second handover request. Different information and parameters can be determined using the value of maximum stay time. This time, and other parameters, can be derived from the invariable offset z with which the MU crosses the cell that it serves. We called this process the *evaluation parameters method* (EPM).

To prove this method works, we used it to reformulate two techniques already present in the literature: last useful instant (LUI) and dynamic channel reservation (DCR) strategies. We called them pseudo-LUI (PLUI) and DCR-*like*.

These two new techniques give very interesting results. In particular, PLUI's performance converged with the performance of LUI. DCR-*like* seemed affected by the locking mechanism used; this is the main cause of its divergence from DCR. Indeed, DCR-*like* has a high probability of new call rejection, but it does considerably improve the probability of handover request being blocked.

The differences we observed are certainly due to the method's delay in determining the required parameters. This delay is due to the absence of all information about ongoing calls before they initialize their second handover request. Though the PLUI strategy is hardly influenced by this delay in information collection, DCR-*like* is greatly influenced. Indeed, if we consider a hexagonal cell for this strategy, the problem gets worse as only the value of maximum stay time in the second transit cell can be used to calculate the probability of call handover.

As a first step, we wanted to show the possibility of using the single mobility model of LEO MSSs to determine some information, even when the exact position of the user is unknown. As a second step, we wanted to improve the EPM by trying to minimize the effects of the constraints we observed; this would enable the proposed systems to perform better.

Indeed, it would be desirable to find a method that enables the determination of values, even if they were approximates, of the maximum queuing time for the central zone of the cells and the maximum stay time in the first transit cell. This would enable these systems to perform better.

In terms of future research, we would suggest taking into consideration the level of signal picked up by the antennae of the two cells (current and transit), in addition to the moment the first handover request initializes in relation to duration of the call since initialization, so as to reach a decision about the position of the MU in relation to the cellular zones.

Making a good decision while considering its different parameters can seem a difficult task. To deal with situations in which some knowledge required for the decision-making is imperfect, fuzzy logic is considered an extremely important and even indispensable tool.

In this context, fuzzy subsets are therefore defined in order to determine the belonging of MUs to different zones in a cell based on the values and information collected.

By defining fuzzy subsets concerning the location of an MU with an ongoing call according to the levels of signal received by the antennae of the source and transit cells, the time between the moment the call initializes and the moment its first handover request initializes, and the time a call spends queuing waiting to be switched to its transit cell, it would be possible to reach a decision about the approximate value of different parameters.

Finally, it would be advantageous to combine the two proposed strategies: PLUI and DCR-*like*.

Second, our study tackled another side of the handover problem: calls that are interrupted a short while after they are initialized. Indeed, we showed that a large number of lost calls had very short durations when they were interrupted. This frustrates users of the network. This problem is worth considering as it is an inconvenience to phone users, particularly when the percentage of these calls is high.

To give calls an equal chance of being served for an acceptable time period, we introduced a decision system based on fuzzy logic, the *rescuing system* (RS), which rescues calls facing a premature end if their length is considered relatively short.

The RS we proposed gave very interesting results: it enabled satellite communication systems to ensure that each call initialized had a guaranteed communication time (1 min in the case we considered) before coming to an abrupt end and considerably improved the average duration of lost calls for different traffic intensity values.

It would be beneficial for satellite communication networks to provide users with a guaranteed period of uninterrupted communication and to improve the average value of the duration of interrupted calls.

To conclude this work, we evaluated the performance of QoS protocols in Internet Protocol (IP) satellite networks using the *DiffServ* architecture. As presented by the IETF, *DiffServ* offers a new approach to QoS that enables differentiated services to be installed in the Internet. The approach was standardized and has advantages that make it stand out from other approaches. New research is currently studying its application in the Internet network as a whole. *DiffServ* assigns the packets entering the network a classification of service (marking); this determines their level of priority and the level of service that they have inside the network. Marking depends on the throughput of ingress traffic, its origin and the state of the network. Nevertheless, congestion can still occur, in particular in the event of torrents of traffic, which requires edge routers to assign ingress packets the lowest level of service, corresponding to *best-effort* service.

QoS demands for satellite networks differ slightly from land networks. However, in terms of QoS, satellite networks come out badly from a comparison with land networks as they have a long propagation delay, a more limited bandwidth and more frequent transmission errors.

Proportional allocation of bandwidth (PAB) presents a new *DiffServ* scheme based on the property of bandwidth allocation during congestion, which is used for geostationary earth orbit (GEO), medium earth orbit (MEO) and low earth orbit (LEO) satellite networks.

PAB will depend, in particular, on the ingress throughput and the committed information rate (CIR) of the flow of packets. It uses labels to show the ratio between the throughput and the CIR of the flow at ingress routers. In the core routers, it uses these labels to drop packets during congestion; there are therefore a number of levels of priority.

What is more, the PAB technique confirms its simplicity and its high performance with regards to delay performance and loss rate, which are in this case "hardly" limited in relation to other techniques.

We conducted several tests on network simulations such as OPNET and NS2. We found that NS2 provided the best possibilities for visualizing IP traffic with proportional allocation of bandwidth.

Appendix 1

Demonstration of Erlang-B and Erlang-C

A1.1. Erlag B

The Erlang B formula is given by:

$$B(S,a) = \frac{\frac{a^S}{S!}}{\sum_{k=0}^{S} \frac{a^k}{k!}}$$

Erlang B for $j-1$ channels is given by:

$$B(j-1,a) = \frac{\frac{a^{j-1}}{(j-1)!}}{\sum_{k=0}^{j-1} \frac{a^k}{k!}}$$

And for j channel by:

$$B(j,a) = \frac{\frac{a^j}{j!}}{\sum_{k=0}^{j} \frac{a^k}{k!}}$$

We have:

$$B(j,a) = \frac{\dfrac{a}{j}\dfrac{a^{j-1}}{(j-1)!}}{\dfrac{a^j}{j!} + \sum_{k=0}^{j-1}\dfrac{a^k}{k!}} = \frac{\dfrac{a}{j}\dfrac{a^{j-1}}{(j-1)!}}{\sum_{k=0}^{j-1}\dfrac{a^k}{k!}\left(\dfrac{a}{j}\dfrac{a^{j-1}}{(j-1)!}\dfrac{1}{\sum_{k=0}^{j-1}\dfrac{a^k}{k!}} + 1\right)}$$

By replacing with the expression of $B(j-1,a)$, we obtain:

$$B(j,a) = \frac{\dfrac{a}{j}B(j-1,a)}{\dfrac{a}{j}B(j-1,a) + 1}$$

and so:

$$B(j,a) = \frac{aB(j-1,a)}{aB(j-1,a) + j}, \qquad (j=1,2,\ldots,S;\ B(0,a)=1)$$

A1.2. Erlang C

The Erlang C formula is given by:

$$C(S,a) = \frac{\dfrac{a^S}{S!(1-a/S)}}{\sum_{k=0}^{S-1}\dfrac{a^k}{k!} + \dfrac{a^S}{S!(1-a/S)}}$$

We can write:

$$C(S,a) = \frac{\dfrac{a^S}{S!(1-a/S)}}{\sum_{k=0}^{S-1}\dfrac{a^k}{k!}\left(1 + \dfrac{a^S}{S!(1-a/S)}\dfrac{1}{\sum_{k=0}^{S-1}\dfrac{a^k}{k!}}\right)}$$

Knowing that:

$$\frac{1}{B(S,a)} = \frac{\sum_{k=0}^{S}\frac{a^k}{k!}}{\frac{a^S}{S!}} = \frac{\frac{a^S}{S!}+\sum_{k=0}^{S-1}\frac{a^k}{k!}}{\frac{a^S}{S!}} = 1 + \frac{\sum_{k=0}^{S-1}\frac{a^k}{k!}}{\frac{a^S}{S!}} = 1 + \frac{S!}{a^S}\sum_{k=0}^{S-1}\frac{a^k}{k!}$$

We have:

$$\Rightarrow \frac{\sum_{k=0}^{S-1}\frac{a^k}{k!}}{\frac{a^S}{S!}} = \frac{1-B(S,a)}{B(S,a)}$$

By replacing in $C(s,a)$:

$$C(S,a) = \frac{\frac{B(S,a)}{1-B(S,a)}\frac{S}{S-a}}{1+\frac{B(S,a)}{1-B(S,a)}\frac{S}{S-a}} = \frac{SB(S,a)}{(1-B(S,a))(S-a)+SB(S,a)}$$

and we obtain:

$$C(S,a) = \frac{SB(S,a)}{S-a+aB(S,a)}$$

Appendix 2

DCA Arrangement

When a call ends, the performance of the network can be improved by using the dynamic channel allocation (DCA) technique, proposed in Chapter 3, which chooses to free a channel based on the deallocation criterion described below.

If $\Lambda(x)$ is the set of channels used in x at the moment a particular channel is freed, we define channel deallocation cost contribution $j \in \Lambda(x)$ due to cell interference $k \in I(x)$, $R_x(k,j)$ as:

$$R_x(k,j) = b_x(k,j) + 2q_k(j), \forall k \in I(x)$$

with b_x (k, j) given by:

$$b_x(k,j) = \begin{cases} 0, \\ 1, \end{cases}$$

if the channel j is closed in k only because of its allocation in x otherwise.

Note that $R_x(k, j)$ can only take four different values:

$$R_x(k,j) = \begin{cases} 0, \text{ if } j \in \Lambda_x(k,j) \text{ and } j \in F_D(k) \\ 1, \text{ if } j \notin \Lambda_x(k,j) \text{ and } j \in F_D(k) \\ 2, \text{ if } j \in \Lambda_x(k,j) \text{ and } j \notin F_D(k) \\ 3, \text{ if } j \notin \Lambda_x(k,j) \text{ and } j \notin F_D(k) \end{cases}$$

where $\Lambda_x(k,j)$ designates the set of channels becoming available in cell k if the channel j is freed in cell x.

By comparing the previous equation and equation [3.6] of Chapter 3, a perfect complementarity between the allocation cost function $C_x(k, j)$ and deallocation cost function $R_x(k, j)$ can be deduced. The main aim of the deallocation algorithm is to reduce divergence with the optimal distribution of fixed channel allocation (FCA) channels and to free the channel that becomes available in the largest number of interference cells.

As such, the total cost function for each channel $j \in \Lambda(x)$ can be obtained by:

$$R_x(j) = 1 - q_x(j) + \sum_{k \in I(x)} \{R_x(k,j)\}, \forall j \in \Lambda(x)$$

The term "$1-q_x(j)$" is introduced in the formula to promote, as much as possible, freeing channels belonging to $F_D(x)$, i.e. the set of channels attributed to x according to FCA.

So, the channel $j^* \in \Lambda(x)$ is freed if it verifies the following relation:

$$R_x(j^*) = \min_{j \in \Lambda(x)} \{R_x(j)\}$$

If the chosen channel j^* differs from the channel that is actually freed $\hat{j}$, the ongoing call in the channel j^* in the cell x must be redirected toward the channel $\hat{j}$.

We have also noticed that $\Lambda(x)$ is updated with each arrival of a new call or freeing of a channel in x, while $\Pi(x)$ is updated with each call arrival or freeing of a channel in x or in $I(x)$.

Appendix 3

Calculating the Probability of Handover Request Failure

We have:

$$P_{b2} = \sum_{n=S}^{+\infty} P_n P_{b2|k}$$

where $k = n\text{-}S$ and S is the number of channels in a cell, given that $P_{b2|k}$ is the probability of a request being put in a queue at the $k+1th$ failed position.

If, in contrast, we take the probability that this request succeeds, it is given by:

$$\left(1 - P_{b2|k}\right) = \left[\prod_{j=1}^{k}\left(P_{j|j+1}\right)\right] P_r$$

Pr is the probability that a channel is freed when the request $k+1$ reaches the top of the queue.

$P_{j|j+1}$ is the probability of the position $j+1$ moving to the position j.

This shift occurs if the call waiting time is higher than: (1) at least one of the call queuing times at one of the previous positions, (2) at least one of the holding times of other calls in a queue at one of the previous positions and (3) at least the ongoing calls' holding times in the transit cell. It is designated by X.

So, we have:

$$1 - P_{j|j+1} = P\{t_{w\,j+1} \leq X\} \times P\{t_{w\,j+1} \leq t_{w1}\}... $$
$$P\{t_{w\,j+1} \leq t_{w\,j}\} \times P\{t_{w\,j+1} \leq t_{o1}\}...P\{t_{w\,j+1} \leq t_{o\,j}\}$$

$t_{w\,j}$ and $t_{o\,j}$ are, respectively, the waiting times and the durations (or holding times) of calls in a queue at positions from 1 to $j+1$.

The probability that $t_{w\,j+1}$ is less than the minimum holding time of ongoing calls in the transit cell is:

$$P_{cl} = \int_0^\infty e^{-S\mu\tau} \mu_\omega e^{-\mu_\omega \tau} d\tau = \frac{\mu_\omega}{S\mu + \mu_\omega}.$$

The probability that $tw\,j+1$ is lower than one of the waiting times of the other calls in a queue Pw is given by:

$$P_w = \int e^{-\mu_\omega \tau} \mu_\omega e^{-\mu_\omega \tau} d\tau = \frac{\mu_\omega}{2\mu_\omega} = \frac{1}{2}$$

and so the probability that $t_{w\,j+1}$ is lower than all the waiting times is:

$$P\{t_{w\,j+1} \leq t_{w\,j}\} P\{t_{w\,j+1} \leq t_{w\,j-1}\}....P\{t_{w\,j+1} \leq t_{w1}\} = \frac{1}{2^j}$$

The probability that $t_{w\,j+1}$ is less than one of the holding times of the other calls in a queue P_o is given by:

$$P_o = \int_0^{+\infty} e^{-\mu\tau} \mu_\omega e^{-\mu_\omega \tau} d\tau = \frac{\mu_\omega}{\mu + \mu_\omega}$$

So, we have:

$$P\{t_{w\,j+1} \leq t_{o\,j}\} P\{t_{w\,j+1} \leq t_{o\,j-1}\} \ldots P\{t_{w\,j+1} \leq t_{o\,1}\} = \left(\frac{\mu_\omega}{\mu + \mu_\omega}\right)^j$$

So, we obtain:

$$1 - P_{j|j+1} = \frac{\mu_\omega}{S\mu + \mu_\omega} \left(\frac{1}{2}\right)^j \left(\frac{\mu_\omega}{\mu + \mu_\omega}\right)^j$$

and then, we obtain:

$$P_{j|j+1} = 1 - \frac{\mu_\omega}{S\mu + \mu_\omega} \left(\frac{\mu_\omega}{\mu + \mu_\omega} \frac{1}{2}\right)^j$$

The probability that no channels in the transit cell are freed when the request $k+1$ is at the top of the queue is given by:

$$P_{cl} = \int_0^\infty e^{-S\mu\tau} \mu_\omega e^{-\mu_\omega \tau} d\tau = \frac{\mu_\omega}{S\mu + \mu_\omega}$$

and so, P_r is equal to $1-P_{cl}$:

$$P_r = \frac{S\mu}{S\mu + \mu_\omega}$$

So, by replacing:

$$1 - P_{b2|k} = \left[\prod_{j=1}^k \left(1 - \frac{\mu_\omega}{S\mu + \mu_\omega} \left(\frac{\mu_\omega}{\mu + \mu_\omega} \frac{1}{2}\right)^j\right)\right] \frac{S\mu}{S\mu + \mu_\omega}$$

we finally obtain:

$$\Rightarrow P_{b2|k} = 1 - \left[\prod_{j=1}^k \left(1 - \frac{\mu_\omega}{S\mu + \mu_\omega} \left(\frac{\mu_\omega}{\mu + \mu_\omega} \frac{1}{2}\right)^j\right)\right] \frac{S\mu}{S\mu + \mu_\omega}$$

So, P_{b2} is equal to:

$$P_{b2} = \sum_{n=S}^{+\infty} P_n \left[1 - \frac{S\mu}{S\mu + \mu_\omega} \left[\prod_{j=1}^{k} \left(1 - \frac{\mu_\omega}{S\mu + \mu_\omega} \left(\frac{\mu_\omega}{\mu_\omega + \mu} \frac{1}{2} \right)^j \right) \right] \right]$$

Appendix 4

Simulation Flow Chart

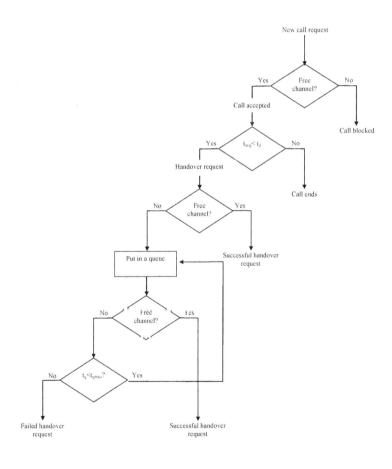

Appendix 5

Presentation of the Network Simulator Software

A5.1. Network simulator

Telecommunication networks, and in particular information technology (IT) networks, have recently experienced unprecedented growth. With the evolution of techniques and technologies, there are countless solutions to a single problem. Simulation enables these new technologies and protocols to be tested as well as the problems that we may face in the future to be predicted. The aim of this appendix is to provide an overview of one of the most used network simulators, Network Simulator (NS), as well as one of the pieces of software that enables simulations to be viewed, Network AniMator (NAM).

A5.1.1. *Introduction*

NS is a simulator of Transmission Control Protocol (TCP)/Internet Protocol (IP) networks. It is the result of a collaboration between the University of Berkeley, Layer by Layer (LBL), University of Southern California/Information Science Institute (USC/ISI) and Xerox PARC through the Virtual Internetwork Tested (VINT) project, supported by Defense Advanced Research Projects Agency (DARP). Using a configuration (nodes and links), network applications can be simulated: constant flow, ftp, telnet and following IP packets throughout their journey. To do this, NS V2 uses a Transmission

Control Link (TCL) language object-oriented version called OTcl, which will be the command and configuration interface through which the user defines the topology of the network, the characteristics of the physical links, the intervening agents or protocols and the communication applications that will occur. Despite the universal use of NS, the documentation is still incomplete and is in no sense progressive. NS is a scalable piece of software: a large number of components are added each year. There are, therefore, countless successive versions of different implemented protocols. Indeed, there are a large number of predefined classes that enable several TCP/IP protocol implementations, different queues, fixed and dynamic routing and faulty links to be implemented.

NS proceeds by discretizing time and produces text files containing the events that took place in the network during simulation. The most interesting use is the animation of the ".nam" files with NAM. This graphic tool interprets the files and gives them a graphic form. It has relatively powerful and unexpected functionalities that, for instance, enable it to quickly provide a plane representation of a complex graph, to follow the loss of packets and to observe the throughput of links.

A5.1.2. *Network simulator: NS*

The development of NS follows an object-oriented approach where two programming languages are used: C++ and TCL (OTcl). The basic modules of the simulator and the protocols are implemented in C++ with an OTcl layer above. The aim of this layer is to provide an interface that is flexible and easy to use. It is possible with NS to simulate a relatively complex IT network containing hundreds of nodes and to be able to view this network with NAM. This information about the progress of the simulation can be retrieved not only with NAM, but also by writing OTcl procedures that enable trace files memorizing the variation of a given parameter in time (such as the number of packets moving in a node, the size of the congestion window of TCP agents, the delay in crossing the network, etc.) to be retrieved. These files can be used by software, such as Xgraph or gnuplot, to obtain the useful curve graphs. Like the majority of freeware, the best documentation on NS is its own source code in C++ and TCL. Nevertheless, a number of tutorials are available on the Internet that explain how NS works. Links to these different tutorials and other additional documentation can be found on the NS website. There is also a reference manual on NS, "NS notes and documentation", which provides more details about the internal workings of NS and presents its different classes; this document is not, however, necessarily up-to-date.

A5.1.2.1. Sources

The executable NS is built from the ns-src.tar.gz archive (or ns-allinone.tar.gz for the all-in-one version). The NS website contains links to various webpages that describe how to install these different components on different platforms. There are two ways to install NS: installing component-by-component, or installing all the components together (the all-in-one version). The first way requires installing TCL, tool kit (TK) and OTcl before installing NS and then NAM; the second way, which appears simpler, has the disadvantage of requiring more memory space (about 100 Mega).

A5.1.2.2. Classes

A5.1.2.2.1. Simulator class

Any ".tcl" file destined to produce a simulation must start by creating a new instance of the class simulator:

set ns [new Simulator]

We can then define the files toward which the two histories *result.nam* and *result.tr* will be directed, which are used by NAM and Xgraph, respectively:

set f1 [open result.nam w]

set f2 [open result.tr w]

$ns namtrace-all $f1

$ns trace-all $f2

Other methods will enable the topology to be created, for example:

set n0 [$ns node] will create a node n0

set n1 [$ns node] will create a node n1

set duplex-link $n0 $n1 2Mb 4ms DropTail will create a full-duplex link between the nodes n0 and n1 with a bandwidth of 2 Mb/s and a propagation time of 4 ms. DropTail shows the discipline of the queue with two extremities, i.e. first in first out (FIFO). There are other disciplines that are also implemented, including Fair Queing (FQ), Weighted Fair Queing (WFQ), Class Based Queing (CBQ), Disaster Risk Reduction (DRR), random early detection (RED), etc. We can also create dissymmetrical links with the instruction "simplexlink".

A5.1.2.2.2. Node class

This class is composed of several other classes, the main one of which is the "classifier" class, which receives each ingress packet and transmits it to the agent linked to the node for which it is destined. There are several types of "classifiers":

– classify according to address "AddressClassifier";

– "Multicast Replicator" classifier;

– "MultiPathForwarder" multipath classifier;

– classify with hash "Hash Classifier";

– the following instruction builds a node N1: set n1 [$ns node].

A5.1.2.2.3. Link class

Links are the secondary pillar for defining the topology of networks. The link class is made up of attributes:

– head_: the ingress point toward the link object;

– queue_: the queue associated with the link;

– link_: refers to the object that models the link based on bandwidth and transmission delay;

– ttl_: refers to the element that handles the ttl in each packet;

– drophead_: refers to the element that handles the dropped packets.

To view dynamic routing, NS includes the notion of faulty links, i.e. links that will be dropped during simulation:

$ns rtmodel-at 1.0 down $n1 $n2 the link between n1 and n2 breaks at instant 1.0.

$ns rtmodel-at 2.0 up $n1 $n2 at the instant 2.0 this link is no longer faulty.

Command for satellite networks:

$ns_ node-config -satNodeType <type>

<type> can be geo, geo-repeater, polar, terminal

$ns_satnode-polar <alt> <inc> <lon> <alpha> <plane> <linkargs> <chan>"

<alt> :altitude   <inc> is orbit inclination   <lon> longitude <alpha> the initial position of the satellite in the orbit

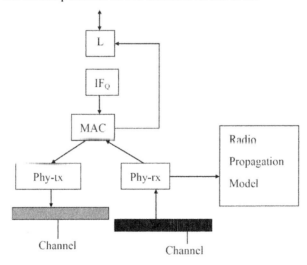

Figure A5.1. *Main component for the interface in satellite networks*

$ns_ node-config -llType <type>

$ns_ node-config -ifqType <type>

$ns_ node-config -ifqLen <length>

$ns_ node-config -macType <type>

$ns_ node-config -channelType <type>

$ns_ node-config -downlinkBW <value>

<linkargs> list of lines defines one network interface (such as Lead Linked (LL), Qtype, Qlim, Physical Layer (PHY), Media Acess Control (MAC), etc).

$ns_ satnode-geo <lon> <linkargs> <chan>

To create a geostationary earth orbit (GEO) satellite with two links (uplink and downlink) and two types of channel:

$node add-interface $type $ll $qtype $qlim $mac $mac_bw $phy

ll: the link layer, qtype: the queue type, qlim: the queue length, mac packets: the MAC, mac_bw: the line bandwidth, phy physical layer

A5.1.2.2.4. Agent class

The agent represents the protocols of the transport layer of the Open System Internetworking (OSI) model. It is responsible for the generation and processing of packets inside a node, so every agent must be attached to a node. As such, Agent/TCP enables a standard TCP entity to be created; and Agent/TCP/Sink is an entity that absorbs the TCP packets that reach it. Agent/constant bit rate (CBR) is another agent that produces traffic with constant throughput and is widely used in demonstrations of NS. It is used with an Agent/Null.

set tcp0 [new Agent/TCP]

set tcp1 [new Agent/TCPSink/DelAck]

$ns attach-agent $n0 $tcp0

$ns attach-agent $n1 $tcp1

To connect the two agents, one sends packets and the other receives them. A connection between these two agents must be defined:

$ns connect $tcp0 $tcp1

To view simulations with NAM, we can mark the packets sent by an agent with a particular color, assuming that we have already defined a color as color 1:

command: $ns color 1 red

$tcp0 set class_ 1

A5.1.2.2.5. Source class

The source class describes the application level. At the moment, only telnet and ftp applications are implemented on NS. These applications must be associated with the agents (the applications will manage the agents). There are, however, agents which do not need to be managed by applications (for instant, the CBR agent: constant bit rate).

set ftp [new source/FTP]	enables a new source to be created
$ftp set agent_ $tcp0	enables this source to be associated with the agent tcp0
$ns at 0.0 "$ftp start"	the source will start sending at the instant 0.0

A5.1.2.2.6. Queue class

The queue is the place where packets will wait to be served. In NS, a number of queue disciplines are implemented: FIFO (DropTail), FQ, WFQ, RED, CBQ, etc. Each of these disciplines processes, i.e. orders and drops, packets differently. The size of the queue can be limited with the following command:

Queue set limit_ lim

A5.1.3. *Network AniMator: NAM*

NAM is the best tool for achieving a graphic representation of the simulation results provided by NS. It has relatively powerful and unexpected functionalities that, for instance, provide us with a plane representation of a complex graph, enable us to follow packet loss, observe the throughputs of different links, etc. Moreover, the visualization of the simulation by NAM can be started, sped up, slowed down, etc. very easily. The use of the zoom enables large networks to be processed. Clicking on a node gives its name. Clicking on a link lets us request the graphic representation of the throughput of this link or the packets it has sent. Clicking on a packet will give information about its type, size, etc. and a marker can be attached to identify it.

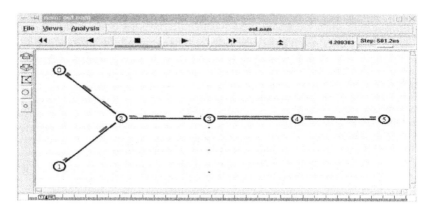

Figure A5.2. *Main window of NAM showing the nodes, links and packets that are circulating*

A5.1.4. *Implementing our approaches with NS*

A5.1.4.1. *The creation of links*

In NS, the *DiffServ* algorithm is implemented on links and not on routers. It is, therefore, necessary to go through simplex links to define the parameters of what is sent and what is received differently:

$ns simplex-link $edge $core 10Mb 5ms dsRED/edge

$ns simplex-link $core $edge 10Mb 5ms dsRED/core

A5.1.4.2. *Types of files*

DiffServ works according to the principle of using virtual queues inside physical queues. We define the priorities, the different characteristics of these queues so that they can be processed as high priority or low priority. We can, notably, vary the probability of dropping a packet:

set qEC [[$ns link $edge $core] queue]

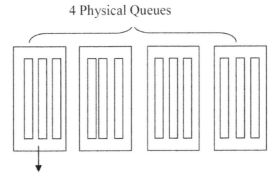

3 Virtual Queues per Physical Queue

Figure A5.3. *Queue class*

A5.1.4.3. *"Edge" link creation function*

To simplify the configuration of "Edge" links, we wrote a TCL function that creates the link configuration. It is worth remembering that it is the responsibility of "Edge" links to make packets and enable the selection of "Core" links:

$qEC meanPktSize $packetSize # packet size

$qEC set numQueues_ 1 # number of physical queues

$qEC set NumPrec 2 # number of virtual queues

$qEC addPolicyEntry [$s1 id] [$dest id] TokenBucket 10 $cir0 $cbs0 #failed packet policy

$qEC addPolicerEntry TokenBucket 10 11 #failed packet policy

$qEC addPHBEntry 10 0 0 #packet 10 in physical queue 0 virtual queue 0

$qEC addPHBEntry 11 0 1 # packet 11 physical queue 0 virtual queue 1

$qEC configQ 0 0 20 40 0.02 # parameter of min queue threshold=20 max threshold=40 drop probability 0.02 for the physical queue 0 virtual queue 0

$qEC configQ 0 1 10 20 0.10 #

Configuration of all queue parameters between the edge and core nodes.

A5.1.4.4. *"Core" link creation function*

The configuration of a "Core" link is more complex than an "Edge" link. First, the same configuration as the "Edge" link must be created such that it knows the different types of flows. Then the "Core" link must be configured by specifying all the parameters for the virtual queues that will help select the packets. This is where we use the majority of the defined queue parameters:

$qCE1 meanPktSize $packetSize

$qCE1 set numQueues_ 1

$qCE1 setNumPrec 2

$qCE1 addPHBEntry 10 0 0

$qCE1 addPHBEntry 11 0 1

$qCE1 configQ 0 0 20 40 0.02

$qCE1 configQ 0 1 10 20 0.10

The queue parameters between core and edge.

Bibliography

[AKR 06] AKRAM A., PERVEZ S., MAJEED N., et al., "Resource management in cellular networks", *Advances in Computer, Information and Systems Sciences and Engineering*, Springer, pp. 89–92, 2006.

[AKY 98] AKYILDIZ I.F., MC NAIR J., HO J., et al., "Mobility management in current and future communications networks", *IEEE Network*, vol. 12, no. 4, pp. 39–49, July-August 1998.

[AKY 99] AKYILDIZ I.F., UZUNALIOGLU H., BENDER M.D., "Handover management in low earth orbit (LEO) satellite networks", *Mobile Networks and Applications*, vol. 4, no. 4, pp. 301–310, December 1999.

[AKY 01a] AKYILDIZ I.F., MORABITO G., PALAZZO S., "TCP- Peach: A new congestion control scheme for satellite IP networks", *IEEE/ACM Transactions on Networking*, vol. 9, no. 3, pp. 307–321, June 2001.

[AKY 01b] AKYILDIZ I.F., MORABITO G., PALAZZO S., "Research Issues for transport protocols in satellite IP networks", *IEEE Personal Communications*, vol. 8, no. 3, pp. 44–48, June 2001.

[AKY 02] AKYILDIZ I.F., EKICI E., BENDER M.D., "MLSR: a novel routing algorithm for multilayered satellite IP Networks", *IEEE/ACM Transactions on Networking*, vol. 10, no. 3, pp. 411–424, June 2002.

[AKY 04] AKYILDIZ I.F., XIE J., MOHANTY S., "A survey of mobility management in next generation all IP based wireless systems", *IEEE Wireless Communications*, vol. 11, no. 4, pp. 16–28, August 2004.

[ALI 97] ALI I., AL DAHIR N., HERSHEY J.E., et al., "Doppler as a new dimension for multiple access in LEO satellite systems", *International Journal of Satellite Communications*, vol. 15, no. 6, pp. 269–279, November-December 1997.

[ALI 99] ALI I., AL DAHIR N., HERSHEY J.E., et al., "Predicting the visibility of LEO satellites", *IEEE Transactions on Aerospace and Electronics Systems*, vol. 35, no. 4, pp. 1183–1190, October 1999.

[ALI 02] ALI I., PIERNO G., BONANI P.G., et al., *Doppler Applications in LEO Satellite Communication Systems*, Kluwer Academic Publishers, 2002.

[ALT 99] ALTMAN E., FERREIRA A., GALTIER J., *Les réseaux satellitaires de télécommunication: Technologie et services*, Dunod, Paris, 1999.

[ARA 05] ARANITI G., IERA A., MOLINARO A., "The Role of HAP in supporting multimedia broadcast and multicast services in terrestrial – satellite integrated systems", *Wireless Personal Communications*, vol. 32, pp. 195–213, 2005.

[AUV 96] AUVRAY G., Portable digital signal transceiver providing communication *via* a terrestrial network and via a satellite network, Patent no. US5564076, 8 October 1996.

[AWE 00] AWEYA J., "On the design of IP routers, Part 1: router architectures", *Journal of Systems Architecture*, vol. 46, pp. 483–511, 2000.

[AWE 04] AWEYA J., MONTUNO D.Y., OUELLETTE M., Method and apparatus for active queue management based on desired queue occupancy, US Patent US 6,690,645 B1, 10 February 2004.

[BAL 80] BALLARD A.H., "Rosette constellations of earth satellites", *IEEE Transactions on Aerospace and Electronic Systems*, vol. 16, no. 5, September 1980.

[BAO 05] BAOWI J., Dynamic frequency selection based on spectrum etiquette, Patent no. US 20080214199 A1, 2005.

[BAY 14] BAYHAN S., ALAGOZ F., "A Markovian approach for best-fit channel selection in cognitive radio networks", *Ad Hoc Networks*, vol. 12, pp. 165–177, 2014.

[BEY 14] BEYME S., LEUNG C., "A stochastic process model of the hop count distribution in wireless sensor networks", *Ad Hoc Networks*, vol. 17, pp. 60–70, 2014.

[BEL 70] BELLMAN R.E., ZADEH L.A., "Decision making in a fuzzy environment", *Management Science*, vol. 17, no. 4, pp. 141–164, 1970.

[BOG 01] BOGGIA G., CAMARDA P., "Modeling dynamic channel allocation in multicellular communication networks", *IEEE Journal on Selected Areas In Communications*, vol. 19, no. 11, pp. 2233–2242, November 2001.

[BOU 95] BOUCHON MEUNIER B., *La logique floue et ses applications*, Editions Addison-Wesley France, 1995.

[BOU 01] BOUKHATEM L., GAITI D., PUJOLLE G., "A channel reservation algorithm for handover issues in LEO satellite systems based on a satellite fixed cell coverage", *IEEE Vehicular Technology Conference*, Atlantic City, NJ, pp. 2975–2979, October 2001.

[BOU 02] BOUKHATEM L., GAITI D., PUJOLLE G., "Resource reservation schemes for handover issue in LEO satellite systems", *Wireless Personal Multimedia Communications, The 5th International Symposium*, pp. 1217–1221, 27-30 October 2002.

[BOU 03a] BOUKHATEM L., BEYLOT A.L., GAITI D., *et al.*, "A time-based reservation scheme for managing handovers in satellite systems", *International Journal of Network Management*, vol. 13, no. 2, pp. 139–145, March-April 2003.

[BOU 03b] BOUKHATEM L., BEYLOT A.L., GAITI D., *et al.*, "TCRA: a time-based channel reservation scheme for handover requests in LEO satellite systems", *International Journal of Satellite Communications and Networking*, vol. 21, no. 3, pp. 227–240, May-June 2003.

[CAO 09] CAO J., STEFANOVIC M., "Switching congestion control for satellite TCP/AQM networks", *American Control Conference*, St Louis, MO, Etats-Unis, pp. 4842–4847, 10-12 June 2009.

[CAP 03] CAPDEROU M., *Satellites Orbits and Missions*, Springer, 2003.

[CHA 01a] CHAN P M.L., SHERIFF R.E., HU Y.F., "An intelligent handover strategy for a multi-segment broadband network", *Personal, Indoor and Mobile Radio Communications, 12th IEEE International Symposium*, vol. 1, pp. C-55–C-59, September 2001.

[CHA 01b] CHAN P.M.L., SHERIFF R.E., HU Y.F., *et al.*, "Mobility management incorporating fuzzy logic for a heterogeneous IP environment", *IEEE Communications Magazine*, vol. 39, no. 12, pp. 42–51, December 2001.

[CHE 96] CHENG M.M.L., CHUANG J.C.I., "Performance evaluation of distributed measurement-based dynamic channel assignment in local wireless communications", *IEEE Journal on Selected Areas in Communications*, vol. 14, no. 4, May 1996.

[CHE 02] CHEN C., EKICI E., AKYILDIZ I.F., "Satellite grouping and routing protocol for LEO/MEO satellite IP networks", *WoWMoM'02*, Atlanta, Etats-Unis, 28 September 2002.

[CHE 05] CHEN C., EKICI E., "A routing protocol for hierarchical LEO/MEO satellite IP networks", *Wireless Networks*, Springer, vol. 11, pp. 507–521, 2005.

[CHE 12] CHENG F., HANG X.F., LEI H.J., "A congestion control scheme for LTE/SAE", *International Conference on Computer Science and Information Technology (ICCSIT 2011) IPCSIT*, IACSIT Press, Singapore, vol. 51, p. 400, 2012.

[CHE 14] CHEN I.-R., GUO J., BAO F., *et al.*, "Trust management in mobile ad hoc networks for bias minimization and application performance maximization", *Ad Hoc Networks*, vol. 19, pp. 59–74, 2014.

[CHO 00] CHO S., AKYILDIZ I.F., BENDER M.D., *et al.*, "New spot beam handover management technique for LEO satellite networks", *IEEE Global Telecommunication Conference*, San Francisco, CA, pp. 1156–1160, 27 November 27-1 December 2000.

[CHO 06a] CHOWDHURY P.K., ATIQUZZAMAN M., IVANCIC W., "Handover schemes in space networks: classification and performance comparison", *2nd IEEE International Conference on Space Mission Challenges for Information Technology (SMC-IT '06)*, Pasadena, CA, pp. 101–108, 17-21 July 2006.

[CHO 06b] CHOWDHURY P.K., ATIQUZZAMAN M., IVANCIC W., "Handover schemes in satellite networks: state-of-the-art and future research directions", *IEEE Communications Surveys and Tutorials*, vol. 8, no. 4, pp. 2–14, August 2006.

[CHO 09] CHOW C.-Y., MOKBEL M.F., NAPS J., *et al.*, "Approximate evaluation of range nearest neighbor queries with quality guarantee", *Advances in Spatial and Temporal Databases*, Lectures notes on Computer Sciences, Springer, vol. 5644, pp. 283–301, 2009.

[CIM 94] CIMINI L.J., FOSCHINI G.J., CHIK LIN I., et al., "Call blocking performance of distributed algorithm for dynamic channel allocation in microcells", *IEEE Transactions on Communications*, vol. 42, no. 8, pp. 2600–2607, 1994.

[CON 88] CONWAY J.H., SLOANE N.J.A., *Sphere Packings; Lattices and Groups*, Springer-Verlag, New York, Berlin, Heidelberg, London, Paris, Tokyo, 1988.

[COO 81] COOPER R.B., *Introduction to Queueing Theory*, North-Holland (Elsevier), New York, 1981.

[COO 90] COOPER R.B., "Queuing theory", *Stochastic Models*, HEYMAN D.P., SOBEL M.J. (eds.), North-Holland (Elsevier), Amsterdam, pp. 469–518, 1990.

[COO 98] COOPER R.B., HEYMAN D.P., "Teletraffic theory and engineering", *Encyclopedia of Telecommunications*, vol. 16, pp. 453–483, 1998.

[DEL 93] DEL RE E., FANTACCI R., GIAMBENE G., "Performance analysis of a dynamic channel allocation technique for terrestrial and satellite mobile cellular networks", *Proceedings Global Telecommunication Conference IEEE GLOBECOM'93*, Houston TX, vol. 3, pp. 1698–1702, 29 November-2 December 1993.

[DEL 94] DEL RE E., FANTACCI R., GIAMBENE G., "Performance analysis of a dynamic channel allocation technique for satellite mobile cellular networks", *International Journal of Satellite Communications*, vol. 12, no. 1, pp. 25–32, January-February 1994.

[DEL 95a] DEL RE E., FANTACCI R., GIAMBENE G., "Efficient dynamic channel allocation techniques with handover queuing for mobile satellite networks", *IEEE Journal Selected Areas in Communications*, vol. 13, pp. 397–405, February 1995.

[DEL 95 b] DEL RE E., FANTACCI R., GIAMBENE G., "Handover and dynamic channel allocation techniques in mobile cellular networks", *IEEE Transactions on Vehicular Technology*, vol. 44, no. 2, pp. 229–237, May 1995.

[DEL 95c] DEL RE E., FANTACCI R., GIAMBENE G., "An efficient technique for dynamically allocating channels in satellite cellular networks", *Proceedings IEEE Global Telecommunications Conference GLOBECOM'95*, Singapore, pp. 1624–1628, 13-17 November 1995.

[DEL 96a] DEL RE E., "A coordinated European effort for the definition of a satellite integrated environment for future mobile communications", *IEEE Communications Magazine*, vol. 34, no. 2, pp. 98–104, February 1996.

[DEL 96 b] DEL RE E., FANTACCI R., RONGA L., "A dynamic channel allocation technique based on hopfield neural networks", *IEEE Transactions on Vehicular Technology*, vol. 45, no. 1, pp. 26–32, February 1996.

[DEL 97] DEL RE E., FANTACCI R., GIAMBENE G., "Performance comparison of different dynamic channel allocation techniques for mobile satellite systems", *European Transactions Telecommunication*, vol. 8, no. 6, pp. 609–621, November-December 1997.

[DEL 99a] DEL RE E., FANTACCI R., GIAMBENE G., "Handover queuing strategies with dynamic and fixed channel allocation techniques in low earth orbit mobile satellite systems", *IEEE Transactions Communications*, vol. 47, no. 1, pp. 89–101, January1999.

[DEL 99 b] DEL RE E., FANTACCI R., GIAMBENE G., "Different queuing policies for handover requests in low earth orbit mobile satellite systems", *IEEE Transactions Vehicular Technology*, vol. 48, no. 2, pp. 448–458, March 1999.

[DEL 08] DEL RE E., RUGGIERI M., "Satellite communication and navigation systems", *Springer Series on Signals and Communication Technology*, pp. 41–50, 2008.

[DEN 01] DENT P.W., Method of directing a call to a mobile telephone in a dual mode cellular satellite network, Patent no. US 61995555, 27 February 2001.

[DIE 96] DIEKELMAN D.P.K., Satellite cellular network resource management method and apparatus, Patent no. US 5590395, 31 December 1996.

[DUR 01] DURRESI A., KOTA S., GOYAL M., *et al.*, "Achieving QoS for TCP traffic in satellite networks with differentiated services", *Space Communications Journal*, vol. 17, nos. 1–3, pp. 125–136, 2001.

[DUR 02] DURESI A., JAGANNATHAN P.K., JAIN R., "Scalable proportional allocation of Bandwidth in IP satellite networks", *IEEE Aerospace conference*, vol. 3, pp.1253–1264, December 2002.

[EFT 98] EFTHYMIOU N., HU Y.F., SHERIFF R.E., *et al.*, "Inter-segment handover algorithm for an integrated terrestrial/satellite-UMTS environment", Personal, Indoor and Mobile Radio Communications, *The 9th IEEE International Symposium PIMRC*, Boston, MA, pp. 993–998, September 1998.

[EL 05] EL ZOOGHBY A., *Smart Antenna Engineering*, Artech House, 2005.

[EIN 01] EINSIEDLER H.J., TELEKOM D., KOLLECKER L., *et al.*, *Testing of Differentiated Services Implementations*, EDIN 0146-P1006 Eurescom, July 2001.

[EKI 02] EKICI E., AKYILDIZ I.F., BENDER M.D., "A multicast routing algorithm for LEO satellite IP networks", *IEEE/ACM Transactions on Networking*, vol. 10, no. 2, pp. 183–192, 2002.

[ETK 07] ETKIN R., PAREKH A., TSE D., "Spectrum sharing for unlicenced bands", *IEEE Journal on Selected Areas in Communications*, vol. 25, no. 3, pp. 517–528, April 2007.

[EVA 05] EVANS B., WERNER M., LUTZ E., *et al.*, "Integration of satellite and terrestrial systems in future multimedia communications", *IEEE Wireless Communications*, vol. 12, no. 5, pp. 72–80, October 2005.

[FAN 04] FAN G., ZHANG J., "A novel geometric diagram and its applications in wireless networks", *INFOCOM Twenty–third annual joint conference of the IEEE Computer and Communications Societies*, vol. 1, pp. 672–682, 2004.

[FER 11] FEROZ A., MUPPALA S.B., OKHOLM J.E., Enhanced randomly early discard for networked devices, US Patent US 2011/0242979 A1, 6 October 2011.

[FLE 06] FLEISCHER F., *Simulation of Typical Modulated Poisson-Voronoi Cells with Applications to Telecommunication Network Modeling*, Frankfurter Stochastik Tage, Frankfurt, 2006.

[GAN 94] GANZ A., YEBIN G., LI B., "Performance study of low earth orbit satellite systems", *IEEE Transactions on Communications*, vol. 42, no. 234, part 3, pp. 1866–1871, February-April 1994.

[GAR 03] GARG S., KAPPES M., "An experimental study of throughput for UDP and VoIP Traffic in IEEE 802.11b Networks", *Wireless Communications and Networking*, vol. 3, pp. 1748–1753, 16–20 March 2003.

[GOY 99] GOYAL R., JAIN R., GOYAL M., et al., "Traffic management for TCP/IP over satellite ATM networks", *IEEE Communications Magazine*, vol. 37, no. 3, pp. 56–61, March 1999.

[GRU 91] GRUBB J.L., "The traveller's dream come true (satellite personal communication)", *IEEE Communications Magazine*, vol. 29, no. 11, pp. 48–51, November 1991.

[GU 07] GU Y., GROSSMAN L., "UDT: UDP-based data transfer for high-speed Wide Area Networks", *Computer Networks*, vol. 51, no. 7, pp. 1777–1799, May 2007.

[GUE 87] GUERIN R., "Channel occupancy time distribution in a cellular radio system", *IEEE Transactions on Vehicular Technology*, vol. 35, no. 3, pp. 89–99, August 1987.

[GUR 91] GURMUNDSON M., "Analysis of handover algorithms", *Proceedings Vehicular Technology Conference, 'Gateway to the Future Technology in Motion', 41st IEEE VTC'91*, pp. 539–543, 1991.

[GUR 07] GURBANI V.K., SUN X.H., "Architecting the telecommunications evolution", *Toward Converged Network Services*, Taylor and Francis Group, 2007.

[HAL 83] HALFIN S., "Batch delays versus customers delays", *Bell System Technical Journal*, vol. 62, no. 7, pp. 2011–2015, September 1983.

[HAS 04] HASSAN M., "High performance TCP/IP networking", *Concepts, Issues, and Solutions*, Pearson Prentice Hall, 2004.

[HEN 99] HENDERSON T.R., KATZ R.H., "Transport protocols for internet compatible satellite networks", *IEEE Journal on Selected Areas in Communications*, vol. 17, no. 2, pp. 326–344, 1999.

[HEY 82] HEYMAN D.P., SOBEL M.J., "Stochastic models in operations research", *Stochastic Processes and Operating Characteristics*, vol. 1, MacGraw Hill, New York, 1982.

[HLU 93] HLUCHYJ M.G., BHARGAVE A., YIN N., Method for prioritizing, selectively discarding, and multiplexing differing traffic type fast packets, US Patent US 5,231,633, 27 February 1993.

[HOM 98] HOMNAN B., BENJAPOLAKUL W., "A handover decision procedure for mobile telephone systems using fuzzy logic", *IEEE APCCAS IEEE Asia-Pacific Conference*, Chiangmai, pp. 503–506, 24-27 November 1998.

[HON 86] HONG D., RAPPAPORT S.S., "Traffic model and performance analysis for cellular mobile radio telephone systems with prioritized and non prioritized handoff procedures", *IEEE Transaction Vehicular Technology*, vol. VT-35, pp. 77–92, August 1986.

[HUA 01] HUANG X.L., BENSAOU B., "On max-min fairness and scheduling in wireless *ad-hoc* networks: analytical framework and Implementation", *MobiHOC*, Long Beach, CVA, USA, 2001

[IBN 04] IBNKAHLA M., RAHMAN Q.M., SULYMAN A.I., *et al.*, "High speed satellite mobile communications: technologies and challenges", *Proceedings of the IEEE*, vol. 92, no. 2, pp. 312–339, February 2004.

[ITU 95] ITU-E-771, Network grade of service parameters and target values for circuit switched land mobile services, Blue Book 1995.

[JAI 99] JAIN R., Design issues for traffic management for the ATM UBR + service for TCP over satellite Networks, NASA/CR-1999-209158, pp. 1–205, July 1999.

[JAM 97] JAMALIPOUR A., *Low Earth Orbit Satellites for Personal Communications Networks*, Artech House, 1997.

[JAM 01] JAMALIPOUR A., TUNG T., "The role of satellites in global IT: trends and implications", *IEEE Personal Communications*, vol. 8, no. 3, pp. 5–11, June 2001.

[JAM 03] JAMALIPOUR A., *The Wireless Mobile internet, Architectures, Protocols and Services*, John Wiley and Sons, 2003.

[JAN 03] JANEVSKI T., *Traffic Analysis and Design of Wireless IP Networks*, Artech House, 2003.

[JUR 07] JURDAK R., *Wireless Adhoc and Sensor Networks, a Cross Layer Design Perspective*, Springer Verlag, 2007.

[KAR 14] KARAHAN A., ERTURK I., ATMACA S, *et al.*, "Effects of transmit-based and receive-based slot allocation strategies on energy efficiency in WSN MACs", *Ad Hoc Networks*, vol. 13, pp. 404 413, 2014.

[KIA 08] KIAMOUCHE W., BENSLAMA M., "Pseudo last useful instant queuing strategy for handovers in low earth orbit mobile satellite networks", *International Journal of Information and Communication Engineering*, vol. 4, no. 5, pp. 369–375, 2008.

[KIA 09] KIAMOUCHE W., BENSLAMA M., "A novel method for evaluating parameters of ongoing calls in low earth orbit mobile satellite systems", *International Journal of Information and Communication Engineering*, vol. 5, no. 5, pp. 344–348, 2009.

[KIA 11a] KIAMOUCHE W., LASMARI S., BENSLAMA M., "Performance analysis of a dynamic channel reservation-like technique for low earth orbit mobile satellite systems", *International Journal of Information and Communication Engineering*, vol. 7, no. 1, pp. 20–25, 2011.

[KIA 11b] KIAMOUCHE W., BENSLAMA M., "Inteligent system rescuing calls with short lengths facing a handover failure in mobile satellite networks", *ICIC Express Letters*, vol. 5, no. 7, pp. 2365–2370, July 2011.

[KIM 96] KIM D.K., SUNG D.K., "Handoff/resource managements based on PCV's and SVC's in broadband personal communication networks", *Proceedings Global Telecommunication Conference IEEE GLOBECOM '96*, London, Royaume-Uni, 18-22 November 1996.

[KIM 03] KIM Y., DiffServ-aware-MPLS: a promising traffic engineering for next generation internet (NGI), Advanced Networking Technology Lab, Korea 2003.

[KOL 02] KOLAWOLE M.O., *Satellite Communications Engineering*, Marcel Deker Editions, New York, 2002.

[KOS 99] KOSTER A., Frequency Assignment – Models and Algorithms, thesis, University of Maastricht, 1999.

[KOT 01] KOTA S., "Quality of Service (QoS) for TCP and UDP traffic over multimedia satellite networks with differentiated services", *AIAA 19th International Communication Satellite Systems*, Toulouse France, 17-20 April 2001.

[KOT 00] KOTA S., DURRESI A., GOYAL M., *et al.*, "A simulation study of Qos for TCP over satellite networks with differentiated services", *Proc. Opnetwork 2000*, August 28-31, pp. 1–16, 2000.

[KÜP 05] KÜPPER A., *Location-based Services Fundamentals and Operation*, John Wiley and Sons, 2005.

[KUS 14] KUSHWAHA V., RATNESHWER, "A review of router based congestion control algorithms", *International Journal of Computer Network and Information Security (IJCNIS)*, vol. 6, no. 1, pp. 1–10, 2013.

[LEE 98] LEE W.C.Y., *Mobile Communications Engineering: Theory and Applications*, McGraw Hill, 1998.

[LI 07] LI M., "Modeling autocorrelation functions of long-range dependent teletraffic series based on optimal approximation in Hilbert space – a further study", *Applied Mathematical Modeling*, vol. 31, pp. 625–631, 2007.

[LIA 02] LIANG Q., "Soft handover for non uniformly-loaded mobile multimedia cellular networks", *Vehicular Technology Conference*, VTC Spring, *IEEE 55th*, pp. 1096–1100, 2002.

[LIL 05] LILITH N., DOGANÇAY K., "Distributed reduced-state SARSA algorithm for dynamic channel allocation in cellular networks featuring traffic mobility", *IEEE International Conference on Communication Seoul Korea ICC 2005*, vol. 2, pp. 860–865, May 2005.

[LIN 94a] LIN Y.-B., CHEN S.C.S., YANG C.S., "Mobility Management for Wireless Systems with Un-reliable Backhaul Links", *IEEE Communications Letters*, vol. 2, no. 5, pp. 122–124, 1998.

[LIN 94b] LIN Y.B., MOHAN S., NOERPEL A., "PCS channel assignment strategies for hand-off and initial access", *IEEE Transactions on Vehicular Technology*, vol. 43, no. 3, p.704–712, 1994.

[LIN 95] LIN Y.-B., CHANG L.F., NOERPEL A., "Modeling hierarchical microcell/macrocell PCS architecture", *Gateway to Globalization, 1995 IEEE International Conference IEEE ICC'95*, Seattle, pp. 405–409, June 1995.

[MAC 79] MACDONALD V.H., "The cellular concept", *Bell System Technical Journal*, vol. 58, pp. 15–41, January 1979.

[MAL 10] MALARVIZHI S., MADHESWARAN M., "New active queue management mechanism for reducing packet loss rate", *Journal of Computer Applications*, vol. III, p. 1, January-March 2010.

[MAR 91] MARAL G., DE RIDDER J.J., EVANS B.G., *et al.*, "Low earth orbit satelllte systems for communications", *International Journal of Satellite Communications*, vol. 9, no. 4, pp. 209–225, 1991.

[MAR 97] MARKOULIDAKIS J.G., TSIRKAS D.F., THEOLOGOU M.E., "The average number of handovers per call in satellite UMTS systems", *Second IEEE Symposium on Computers and Communications (ISCC'97)*, Alexandria, Egypt, pp. 370–374, 1-3 July 1997.

[MAR 98] MARAL G., RESTREPO J., DEL RE E., et al., "Performance analysis for a guaranteed handover service in an LEO constellation with a 'satellite-fixed cell' system", *IEEE Transactions on Vehicular Technology*, vol. 47, no. 4, pp. 1200–1214, November 1998.

[MAT 07] MATHIOUDAKIS I., WHITE N.M., HARRIS N.R., "Wireless sensor networks: applications utilizing satellite links", *18th Annual IEEE International Symposium on Personal, Indoor and Mobile Radio communications (PIMRC'07)*, pp. 1–5, 2007.

[MCC 07] MC CABE J.D., *Network Analysis, Architecture and Design*, 3rd ed., Mogan Kaufmann, June 2007.

[MCN 04] MCNAIR J., ZHU F., "Vertical handoffs in fourth – generation multinetwork environments", *IEEE Wireless Communications*, vol. 11, no. 3, pp. 8–15, June 2004.

[MIS 01a] MISRA A., OTT T., BARAS J., "Effect of Exponential averaging on the variability of a RED queue", *IEEE International Conference on Communications ICC 2001*, vol. 6, pp. 1817–1823, 2001.

[MIS 03b] MISRA A., OTT T., BARAS J., "Markov Processes with State-Dependent Failure Rates and Applications to RED and TCP windows dynamics", *11th Mediterranean Conference on Control and Automation, (MED'03)*, Rhodes, Greece, June 18-20, 2003.

[MIS 02] MISRA A., OTT T., BARAS J., "Predicting bottleneck bandwidth sharing by generalized flows", *Computer Networks*, vol. 40, pp. 557–576, 2002.

[MON 05] MONTENBRUCK O., GILL E., *Satellite Orbits, Models, Methods, Applications*, Springer, 2005.

[NIC 03] NICHOLS K.M., Method and apparatus for providing differentiated services using a multi-level queuing mechanism, US Patent 6,608,816 B1, 19 August 2003.

[NIE 06] NIE N., COMANICIU C., "Adaptive channel allocation spectrum etiquette for cognitive radio networks", *Journal Mobile Networks and Applications*, vol. 11, no. 6, pp. 779–797, December 2006.

[OBR 99] OBRADOVIC V., CIGOJ S., "Performance evaluation of prioritized handover management for LEO mobile satellite systems with dynamic channel assignment", *IEEE Global Telecommunication Conference*, Rio de Janeiro, Brazil, pp. 296–300, December 1999.

[OH 07] OH D.G., PANSOO K., SONG J.J., *et al.*, "Design considerations of satellite based vehicular broadband networks", *IEEE Wireless Communications*, vol. 12, no. 5, pp. 91–97, October 2005.

[OUE 07] OUELLETE M., AYEWA J., DASYLVA A., *et al.*, Queue based multilevel AQM with drop precedence differentiation, US Patent 7,286,485 B1, 23 October 2007.

[PAL 95] PALLANT D.L., TAYLOR P.G., "Modeling handover in cellular mobile networks with dynamic channel allocation", *Operations Research*, vol. 43, no. 1, pp. 33–42, January-February 1995.

[PAP 03] PAPAPETROU E., PAVLIDOU F.N., "QoS handover management in LEO/MEO satellite systems", *Wireless Personal Communications*, vol. 24, no. 2, pp. 189–204, February 2003.

[PAP 04 a] PAPAPETROU E., KARAPANTAZIS S., DIMITRIADIS G., *et al.*, "Satellite handover techniques for LEO networks", *International Journal of Satellite Communications and Networking*, vol. 22, no. 2, pp. 231–245, March-April 2004.

[PAP 04 b] PAPAPETROU E., KARAPANTAZIS S., PAVLIDOU F.N., "Distributed on demand routing for LEO satellite systems", *Computer Networks*, vol. 512, no. 15, pp. 4356–4376, October 2007.

[PAR 04] PARK E.C., CHOI C.H., "Proportional bandwidth in diffserv networks", *IEEE Infocom Twenty third annual joint conference of the IEEE Computer and Communications Societies*, vol. 3, pp. 2038–2049, 2004.

[PAT 05] PATEL P.R., PATEL A.R., CHOLSKI O.T., *et al.*, Method and system for queuing traffic in a wireless communications network, US Patent US 6,865,185 B1, 8 March 2005.

[PEN 14] PENG S., LOW C.P., "Prediction free energy neutral power management for energy harvesting wireless sensor nodes", *Ad Hoc Networks*, vol. 13, pp. 351–367, 2014.

[PRA 05] PRASAD R., RUGGIERI M., *Applied Satellite Navigation Using GPS, Galileo, and Augmentation System*, Artech House, 2005.

[PUR 95] PURZYNSKI C., RAPPAPORT S.S., "Multiple call hand-off problems with queued hand-off and mixed platform types", *Proceedings Communications IEEE*, vol. 142, no. 1, part I, pp. 31–39, February 1995.

[QUI 04] QUILES B.P., An enhanced mobility model, thesis, Technischen Universität Wien, March 2004.

[RAO 14] RAO Y., YUAN C.A., JIANG Z.H., et al., "Agent based multi-service routing for polar – orbit LEO broadband satellite networks", *Ad Hoc Networks*, vol. 13, pp. 575–597, 2014.

[RAP 93] RAPPAPORT S.S., "Blocking, hand-off and traffic performance for cellular communication systems with mixed platforms", *IEEE Proceedings Communications, Speech and Vision*, vol. 140, no. 5, part I, pp. 389–401, October 1993.

[RAY 91] RAYMOND P.A., "Performance analysis of Cellular Networks", *IEEE Transactions on Communications*, vol. 39, no. 12, pp. 1787–1793, December 1991.

[RAY 03] RAYCHAUDHURI D., JING X., "A spectrum etiquette protocol for efficient coordination of radio devices in unlicensed bands", *PIMRC, IEEE proceedings*, vol. 1, pp. 172–176, 7-10 September 2003.

[RES 95] RESTREPO J., MARAL G., "Coverage concepts for satellite constellations providing communications services to fixed and mobile users", *Space Communications*, vol. 13, no. 2, pp. 145–157, 1995.

[ROS 83] ROSS S.M., *Stochastic Processes*, Wiley, New York, 1983.

[ROS 93] ROSS S.M., *Introduction to Probability Models*, 5th ed., Academic Press, San Diego, CA, 1993.

[ROS 95] ROSS K.W., *Multiservice Loss Models for Broadband Telecommunication Networks*, Springer-Verlag, New York, 1995.

[ROS 99] ROSOLEN V., BONAVENTURE O., LEDUC G., "A RED discard strategy for ATM networks and its performance evaluation with TCP/IP traffic", *ACM. SIGCOMM. Computer Communications Review*, vol. 29, no. 3, pp. 23–43, July 1999.

[RYB 05] RYBKO A.N., SHLOSMAN S.B., "Poisson hypothesis", *Journal Problems of Information Transmission*, vol. 41, no. 3, pp. 230–236, July 2005.

[SAA 61] SAATY T.L., *Elements of Queuing Theory*, Mac Graw Hill, 1961.

[SAH 98] SAHA M.K., "Interfering cell identification in satellite cellular system", *IEEE Transactions on Aerospace and Electronics Systems*, vol. 34, no. 2, pp. 477–485, April 1998.

[SAN 95] SANTOS V., SILVA R., DINIS M., et al., "Performance evaluation of channel assignment strategies and handover policies for satellite mobile networks", *Annual International Conference on Universal Personal Communications*, Tokyo, Japan, pp. 86–90, November 1995.

[SAN 08] SANTIAGO R.C., et al., "Enhanced efficiency and frequency assignment by optimizing the base stations location in a mobile radio network", *Wireless Networks*, vol. 14, pp. 531–541, 2008.

[SEL 06] SELIGMAN M., FALL K., MUNDUR P., "Alternative custodians for congestion control in delay tolerant networks", *SIGCOMM'06 Workshops*, Pisa, Italy, 11-15 September 2006.

[SHE 96] SHEIKH A.U., MLONJA C.H., "Performance of fuzzy algorithm based handover process for personal communication systems", *Proceedings of Personal Wireless Communications, IEEE International Conference*, New Delhi, India, 19-21 February, 1996.

[SHE 98] SHERIFF R.E., HU Y.F., DEL RE E., et al., "Satellite-UMTS traffic dimensioning and resource management technique analysis", *IEEE Transactions Vehicular Technology*, vol. 47, no. 4, pp. 1329–1341, November 1998.

[SHE 01] SHERIFF R.E., FUN HU Y., *Mobile Satellite Communication Networks*, John Wiley & Sons Ltd, 2001.

[SUN 05] SUN Z., *Satellite Networking, Principles and Protocols*, John Wiley & Sons, 2005.

[SOR 99] SORACE R., "Overview of multiple satellite communication networks", *IEEE Transactions on Aerospace and Electronics Systems*, vol. 35, no. 4, pp. 1362–1368, October 1999.

[STE 94] STEWART W.J., *Introduction to the Numerical Solution of Markov Chain*, PUP Princeton, 1994.

[STO 98] STOICA I., SHENKER S., ZHANG H., "Core-stateless fair queuing: achieving approximately fair bandwidth allocations in high speed networks", *Computer Communication Review*, vol. 28, no. 4, pp. 118–130, October 1998.

[SYS 86] SYSKI T., *Introduction to Congestion Theory in Telephone Systems*, 2nd ed., Elsevier, New York, 1986.

[TAL 05] TALEB T., KATO N., NEMOTO Y., "Recents trends in IP/NGEO satellite communications systems: transport, routing and mobility management concerns", *IEEE Wireless Communications*, vol. 12, no. 5, pp. 63–69, October 2005.

[TAL 06] TALEB T., KATO N., NEMOTO Y., "REFWA: An efficient and fair congestion control scheme for LEO Satellite Networks", *IEEE/ACM Transactions on Networking*, vol. 14, no. 5, pp. 1031–1044, October 2006.

[TAL 09] TALEB T., MASHIMO D., JAMALIPOUR A., *et al.*, "Explicit load balanced technique for NGEO satellite IP networks on board processing capabilities", *IEEE/ACM Transactions on Networking*, vol. 17, no. 1, pp. 281–293, February 2009.

[TEK 91] TEKINAY S., JEBBARI B., "Handover and channel assignment in mobile cellular networks", *IEEE Communications Magazine*, vol. 29, no. 11, pp. 42–46, November 1991.

[TEK 92a] TEKINAY S., JABBARI B., "A measurement-based prioritization scheme for handovers in mobile cellular networks", *IEEE Journal Selected Areas in Communications*, vol. 10, pp. 1343–1350, October 1992.

[TEK 92b] TEKINAY S., JABBARI B., "Analysis of measurement based prioritization schemes for handovers in cellular networks", *Proceedings Global Telecommunication Conference, IEEE GLOBECOM'92*, Orlando, FL, 1992.

[TIA 01] TIAN X., JI C., "Bounding the performance of dynamic channel allocation with QoS provisioning for distributed admission control in wireless networks", *IEEE Transactions on Vehicular Technology*, vol. 50, no. 2, pp. 388–397, March 2001.

[TSE 07] TSENG Y.M., "A heterogeneous – network aided public – key management scheme for mobile *ad hoc* networks", *International Journal of Network Management*, vol. 17, pp. 3–15, 2007.

[TSU 04] TSUNODA H., OHTA K, KATO N., *et al.*, "Supporting IP/LEO satellite networks by handover-independent IP mobility management", *IEEE Journal on Selected Areas in Communications*, vol. 22, no. 2, pp. 300–307, February 2004.

[UYA 14] UYANIK G.S., ABDERAHMAN M.J., KRUNZ M., "Optimal channel assignment with aggregation in multi-channel systems: a resilient approach to adjacent-channel interference", *Ad Hoc Networks*, vol. 20, pp. 64–76, 2014.

[VAR 03] VARSHNEY U., "Location management for mobile commerce applications in wireless internet environnement", *ACM Transactions on Internet Technology*, vol. 3, pp. 236–255, August 2003.

[VAL 07] VALLAMSUNDAR B., ZHU J., PONNAMBALAM K., et al., "Congestion control for adaptive satellite communication systems using intelligent systems", *ISSSE'07, International Symposium on Signals, Systems and Electronics*, pp. 415–418, July 30-August 2, 2007.

[WAL 71] WALKER J.G., "Some circular orbit patterns providing continuous whole earth coverage", *Journal of the British Interplanetary Society*, vol. 24, pp. 369–384, 1971.

[WAL 84] WALKER J.G., "Satellite constellations", *Journal of the British Interplanetary Society*, vol. 37, pp. 559–571, 1984.

[WAL 07] WALLACE K., CCVP QOS quick Reference sheets, Cisco Press, 2007.

[WAN 93] WANG J.Z., "Simulation and performance analysis of dynamic channel allocation algorithms", *DECT' IEEE Transactions on Vehicular Technology*, vol. 42, no. 4, pp. 563–569, November 1993.

[WAN 01] WANG Z., MATHIOPOULOS P.T., "Analysis and performance evaluation of dynamic channel reservation techniques for LEO mobile satellite systems", *IEEE Vehicular Technology Conference VTC Spring*, VTS 53rd, vol. 4, pp. 2985–2989, 2001.

[WAN 02] WANG Z., MATHIOPOULOS P.T., "A novel traffic dependant dynamic channel allocation and reservation technique for LEO mobile satellite systems", *Proceedings IEEE Vehicular Technology Conference 56th VTC '02*, vol. 3, pp. 1652–1656, 2002.

[WAN 07] WANG L., SHEN H., CHEN Z., et al., "Voronoi tessellation based rapid coverage decision algorithm for wireless sensor networks", INDULSKA J., et al., (eds.), *UIC 2007, LNCS 4611*, Springer Verlag, Berlin Heidelberg, pp. 495–502, 2007.

[WAN 14] WANG J., ZHU D., ZHANG H., et al., "Resource optimization for cellular network assisted multichannel D2D communication", *Signal Processing*, vol. 100, pp. 23–31, 2014.

[WOL] WOLF R.W., *Stochastic Modeling and the Theory of Queues*, Prentice Hall, Englewood Cliffs, NJ, 1989.

[WON 00] WONG V.W.S., LEUNG V.C.M., "Location management for next generation personal communications networks", *IEEE Network*, vol. 14, no. 5, pp. 18–24, September –October 2000.

[WYS 05] WYSOCKI T.A., DADEJ A., WYSOCKI B.J., *Advanced Wired and Wireless Networks*, Springer, 2005.

[XIE 07] XIE W., WANG R., CAO W., "Analysis of higher order Voronoi diagram for fuzzy information coverage", ZHANG H., *et al.* (eds.), *MSN 2007*, LNCS 4864, Springer Verlag, Berlin Heidelberg, pp. 616–622, 2007.

[XIE 12] XIE Z., MA L., LIANG X., "Unlicensed spectrum sharing game between LEO satellites and terrestrial cognitive radio networks", *Chinese Journal of Aeronautics*, vol. 25, pp. 605–614, 2012.

[XU 00] XU Y., DING Q., KO C.C., "Elastic handover scheme for LEO satellite mobile communication systems", *IEEE Global Telecommunication Conference*, San Francisco, CA, pp. 1161–1165, 27 November-1er December, 2000.

[XU 04] XU K., TIAN Y., ANSARI N., "TCP-Jersey for Wireless IP communications", *IEEE Journal on Selected Areas in Communications*, vol. 22, no. 4, p. 747, May 2004.

[YAC 02] YACOUB M.D., *Wireless Technology, Protocols, Standards and Techniques*, CRC Press LLC, 2002.

[ZAD 65] ZADEH L.A., "Fuzzy sets", *Information and Control Elsevier*, vol. 8, no. 3, pp. 338–353, June 1965.

[ZAH 02] ZAHARIADIS T.B., VAXEVANAKIS K.G., TSANTILAS C.P., *et al.*, "Global roaming in next-generation networks", *IEEE Communications Magazine*, p. 145, February 2002.

[ZEN 07] ZENG H., SRINIVASAN A., CHENG B., *et al.*, "Adaptive congestion control under dynamic weather condition for wireless and satellite networks", *ITC 2007*, LNCS 4516 Springer, pp. 92–103, 2007.

[ZHA 89a] ZHANG M., YUM T.S.P., "Comparisons of channel assignment strategies in cellular mobile telephone systems", *IEEE Transactions on Vehicular Technology*, vol. 38, no. 4, pp. 211, November 1989.

[ZHA 89b] ZHANG M., YUM T.S., "Comparison of channel assignment strategies in cellular mobile telephone systems", *IEEE Transactions on Vehicular Technology*, vol. 38, no. 4, pp. 211–215, November 1989.

[ZHA 95a] ZHAO W., TAFAZOLLI R., EVANS B.G., "A UT positioning approach for dynamic satellite constellations", *Proceedings International Mobile Satellite Conference 4th, IMSC'95*, Ottawa, Canada, pp. 251–258, 6-8 June 1995.

[ZHA 95b] ZHANG S., BURNS A., "An optimal synchronous bandwidth allocation scheme for guaranteeing synchronous message deadlines with the timed MAC protocol", *IEEE/ACM Transactions on Networking*, vol. 3, no. 6, p. 729, December 1995.

[ZHA 03] ZHANG Y., *Internetworking and Computing over Satellite Networks*, Kluwer Academic Press, Mars 2003.

[ZHA 04a] ZHANG W., "Handover decision using fuzzy MADM in heterogeneous networks", *IEEE Wireless Communications and Networking Conference (WCNC '04)*, Atlanta, GA, 21-25 March 2004.

[ZHA 04b] ZHANG Y., "A multilayer IP security protocol for TCP performance enhancement in wireless Networks", *IEEE Journal on Selected Areas in Communications*, vol. 22, pp. 767–776, May 2004.

[ZIN 14] ZINONOS Z., CHRYSOSTOUMOU C., VASSILIOU V., "Wireless sensor networks mobility management using fuzzy logic", *Ad Hoc Networks*, vol. 16, pp. 70–87, 2014.

Index

B, C, D, E

birth-death process, 27, 31–33
channel allocation strategy, 39
DCR technique, 84
ellipse, 3, 4, 6
elliptical systems, 15, 16
Erlang-B model, 34–37
Erlang-C model, 34–37
evaluation parameters
 method, 63

F, G

FCA-QH, 85, 86, 90, 91, 113–115
FCR(-like), 85, 91–100, 119, 120
FIFO strategy, 79, 81, 88, 89, 118
fixed channel reservation (FCR), 85
fuzzy logic, 101–105, 153, 154
GEO-type systems, 14–15
guard channel
 strategy, 81–84

I, K, L

IP traffic, 127
Kepler's laws, 4, 7, 15
LEO MSS mobility level, 64
LEO-type systems, 17
LUI queuing strategy, 78–79, 85, 91, 113

M, O, P

MEO-type systems, 17
mobile users (MU), 46–55, 60, 63–73, 77, 84, 90, 103, 122, 152, 153
mobility model, 39, 40, 48–53, 63–65, 68, 69, 71, 74, 79, 81, 92, 103, 112, 119, 122
orbital
 parameters, 5
 paths, 13–20
 perturbations, 7
packet labeling
 methodology, 131–133
PLUI strategy, 79, 80, 87–90, 114, 119, 152

Poisson process, 27, 32–34, 36, 86, 92, 94, 95, 112
proportional allocation of bandwidth (PAB), 127
pseudo-LUI strategy, 64, 79–81, 85

Q, R, S, T

queuing theory, 27–30
rescuing system, 105, 110, 154
satellite networks, 1, 127
satellite orbits, 3–8
spotbeam handover, 24, 39, 43–48, 57, 101
teletraffic theory, 27–30, 32

Other titles from

in

Networks and Telecommunications

2014

BATTU Daniel
New Telecom Networks: Enterprises and Security

BEN MAHMOUD Mohamed Slim, GUERBER Christophe, LARRIEU Nicolas, PIROVANO Alain, RADZIK José
Aeronautical Air−Ground Data Link Communications

BITAM Salim, MELLOUK Abdelhamid
Bio-inspired Routing Protocols for Vehicular Ad-Hoc Networks

CAMPISTA Miguel Elias Mitre, RUBINSTEIN Marcelo Gonçalves
Advanced Routing Protocols for Wireless Networks

CHETTO Maryline
Real-time Systems Scheduling 1: Fundamentals
Real-time Systems Scheduling 2: Focuses

EXPOSITO Ernesto, DIOP Codé
Smart SOA Platforms in Cloud Computing Architectures

MELLOUK Abdelhamid, CUADRA-SANCHEZ Antonio
Quality of Experience Engineering for Customer Added Value Services

OTEAFY Sharief M.A., HASSANEIN Hossam S.
Dynamic Wireless Sensor Networks

PEREZ André
Network Security

PERRET Etienne
Radio Frequency Identification and Sensors: From RFID to Chipless RFID

REMY Jean-Gabriel, LETAMENDIA Charlotte
LTE Standards
LTE Services

TANWIR Savera, PERROS Harry
VBR Video Traffic Models

VAN METER Rodney
Quantum Networking

XIONG Kaiqi
Resource Optimization and Security for Cloud Services

2013

ASSING Dominique, CALÉ Stéphane
Mobile Access Safety: Beyond BYOD

BEN MAHMOUD Mohamed Slim, LARRIEU Nicolas, PIROVANO Alain
Risk Propagation Assessment for Network Security: Application to Airport Communication Network Design

BEYLOT André-Luc, LABIOD Houda
Vehicular Networks: Models and Algorithms

BRITO Gabriel M., VELLOSO Pedro Braconnot, MORAES Igor M.
Information-Centric Networks: A New Paradigm for the Internet

BERTIN Emmanuel, CRESPI Noël
Architecture and Governance for Communication Services

DEUFF Dominique, COSQUER Mathilde
User-Centered Agile Method

DUARTE Otto Carlos, PUJOLLE Guy
Virtual Networks: Pluralistic Approach for the Next Generation of Internet

FOWLER Scott A., MELLOUK Abdelhamid, YAMADA Naomi
LTE-Advanced DRX Mechanism for Power Saving

JOBERT Sébastien *et al.*
Synchronous Ethernet and IEEE 1588 in Telecoms: Next Generation Synchronization Networks

MELLOUK Abdelhamid, HOCEINI Said, TRAN Hai Anh
Quality-of-Experience for Multimedia: Application to Content Delivery Network Architecture

NAIT-SIDI-MOH Ahmed, BAKHOUYA Mohamed, GABER Jaafar, WACK Maxime
Geopositioning and Mobility

PEREZ André
Voice over LTE: EPS and IMS Networks

2012

AL AGHA Khaldoun
Network Coding

BOUCHET Olivier
Wireless Optical Communications

DECREUSEFOND Laurent, MOYAL Pascal
Stochastic Modeling and Analysis of Telecoms Networks

DUFOUR Jean-Yves
Intelligent Video Surveillance Systems

EXPOSITO Ernesto
Advanced Transport Protocols: Designing the Next Generation

JUMIRA Oswald, ZEADALLY Sherali
Energy Efficiency in Wireless Networks

KRIEF Francine
Green Networking

PEREZ André
Mobile Networks Architecture

2011

BONALD Thomas, FEUILLET Mathieu
Network Performance Analysis

CARBOU Romain, DIAZ Michel, EXPOSITO Ernesto, ROMAN Rodrigo
Digital Home Networking

CHABANNE Hervé, URIEN Pascal, SUSINI Jean-Ferdinand
RFID and the Internet of Things

GARDUNO David, DIAZ Michel
Communicating Systems with UML 2: Modeling and Analysis of Network Protocols

LAHEURTE Jean-Marc
Compact Antennas for Wireless Communications and Terminals: Theory and Design

RÉMY Jean-Gabriel, LETAMENDIA Charlotte
Home Area Networks and IPTV

PALICOT Jacques
Radio Engineering: From Software Radio to Cognitive Radio

PEREZ André
IP, Ethernet and MPLS Networks: Resource and Fault Management

TOUTAIN Laurent, MINABURO Ana
Local Networks and the Internet: From Protocols to Interconnection

2010

CHAOUCHI Hakima
The Internet of Things

FRIKHA Mounir
Ad Hoc Networks: Routing, QoS and Optimization

KRIEF Francine
Communicating Embedded Systems / Network Applications

2009

CHAOUCHI Hakima, MAKNAVICIUS Maryline
Wireless and Mobile Network Security

VIVIER Emmanuelle
Radio Resources Management in WiMAX

2008

CHADUC Jean-Marc, POGOREL Gérard
The Radio Spectrum

GAÏTI Dominique
Autonomic Networks

LABIOD Houda
Wireless Ad Hoc and Sensor Networks

LECOY Pierre
Fiber-optic Communications

MELLOUK Abdelhamid
End-to-End Quality of Service Engineering in Next Generation Heterogeneous Networks

PAGANI Pascal *et al.*
Ultra-wideband Radio Propagation Channel

2007

BENSLIMANE Abderrahim
Multimedia Multicast on the Internet

PUJOLLE Guy
Management, Control and Evolution of IP Networks

SANCHEZ Javier, THIOUNE Mamadou
UMTS

VIVIER Guillaume
Reconfigurable Mobile Radio Systems